# Making the Move to Cross Docking

*A practical guide to planning, designing, and implementing a cross dock operation*

by **Maida Napolitano**
and the staff at **Gross & Associates**

**WERC**

Warehousing Education and Research Council
1100 Jorie Boulevard, Suite 170
Oak Brook, IL 60523-4413

| | |
|---|---|
| TEL | 630-990-0001 |
| FAX | 630-990-0256 |
| WEB | www.werc.org |
| E-MAIL | wercoffice@werc.org |

Gross & Associates is an independent management consulting firm specializing in logistics, physical distribution, facilities design, and material handling. Over the past 38-plus years, Gross & Associates has designed thousands of warehousing/distribution operations in the United States, Canada, Mexico, South America, Europe, and Asia. The firm's clients include many Fortune 1000 corporations as well as smaller companies. Clients range from those requiring highly automated material handling systems design and sophisticated warehouse management systems support to those requiring less-mechanized productivity solutions.

Copyright ©2000, Warehousing Education and Research Council. All rights reserved. No part of this publication may be copied or reproduced in any manner without the express written permission of WERC's Executive Director.

ISBN 1-892663-18-X

# Table of Contents

*List of Figures*  v

*Acknowledgments*  vii

*Foreword*  ix

*How to Use This Book*  xi

**CHAPTER 1**  **Cross Docking at a Glance**  1
- **1.1**  Defining cross docking  6
- **1.2**  A brief history of cross docking  8
- **1.3**  Types of cross docking applications  9
- **1.4**  Ideal conditions for success  14
- **1.5**  Cross docking cannot exist alone  16
- **1.6**  Choosing distribution strategies  17
- **1.7**  Benefits and drawbacks of cross docking  19
- **1.8**  Applying cross docking  20
- **1.9**  Trends that stress the importance of cross docking  21
- **1.10**  CD is here to stay  26

**CHAPTER 2**  **Phase 1: Assessing the Potential for Cross Docking**  27
- **2.1**  Step 1: Review current corporate goals and trends  29
- **2.2**  Step 2a: Review products  32
- **2.3**  Step 2b: Review and select suppliers  42
- **2.4**  Step 2c: Audit current operations and policies  49
- **2.5**  Step 3: Analyze strengths and weaknesses  52
- **2.6**  Step 4: Develop short- and long-term recommendations  59
- **2.7**  Step 5: Quantify recommendations  60
- **2.8**  Step 6: Review recommendations with appropriate trading partners and negotiate the benefits and costs  62
- **2.9**  Step 7: Finalize recommendations  66

**CHAPTER 3**  **Phase 2: Planning and Designing a Cross Docking System**  67
- **3.1**  Types of cross docking systems  68
- **3.2**  Step 1: Generate cross docking system designs  91
- **3.3**  Step 2: Perform an economic analysis on the alternatives  96

| | | |
|---|---|---|
| **CHAPTER 4** | | **Phase 3: Identifying Costs and Savings of a Cross Docking System** 105 |
| | 4.1 | Step 1: Create a system-wide cost model and determine ROI 107 |
| | 4.2 | Step 2: Create a product cost model and determine impact on product profitability 116 |
| | 4.3 | Step 3: Calculate the exact costs and savings 121 |
| **CHAPTER 5** | | **Phase 4: Implementing and Maintaining the Cross Docking System** 123 |
| | 5.1 | Step 1: Form a cross-functional team and verify objectives 125 |
| | 5.2 | Step 2: Locate cross dock center and select site (only for new facilities) 126 |
| | 5.3 | Step 3: Develop an implementation schedule 127 |
| | 5.4 | Step 4: Detail plans of required changes 128 |
| | 5.5 | Step 5: Train personnel 134 |
| | 5.6 | Step 6: Procure equipment 134 |
| | 5.7 | Step 7: Procure or upgrade information systems 136 |
| | 5.8 | Step 8: Prepare cross dock site 137 |
| | 5.9 | Step 9: Implement a pilot program 137 |
| | 5.10 | Step 10: Implement cross docking on a system-wide basis 139 |
| | 5.11 | Step 11: Periodically review cross docking operation 139 |
| | 5.12 | Step 12: Consider improving and expanding the cross dock program 140 |
| | 5.13 | Pitfalls to avoid and lessons learned 140 |
| **CHAPTER 6** | | **A Case Study On Cross Docking** 147 |
| | 6.1 | Background 148 |
| | 6.2 | Generating cross dock system designs 149 |
| | 6.3 | Economic analysis of alternatives 158 |
| | 6.4 | Implementation 161 |
| | 6.5 | One year of operation 161 |
| | 6.6 | The future for S&O 162 |
| **APPENDIX A** | | **Organizations and Information Sources for Cross Docking** 165 |
| **APPENDIX B** | | **General Dock Design** 169 |
| | | Design Year Requirements 169 |
| | | Dock Design 170 |
| | | *Bibliography* 175 |
| | | *Index* 179 |

# List of Figures

1. Defining Cross Docking within a Facility   6
2. Generic Definition of Cross Dock Players   7
3. How Cross Docking Supports Just-In-Time (JIT)   10
4. Post-Production Cross Docking   11
5. Elements for the Ideal Cross Dock Program   15
6. Different Distribution Strategies   17
7. Cross Dock Savings Based on Grocery Pilot Programs   18
8. Cross Docking Benefits Versus Drawbacks   19
9. Making the Transition to Cross Docking   20
10. Illustration of Continuous Replenishment with Cross Docking   23
11. Phase 1: Assessment Process   29
12. Phase 1, Step 2a: Product Review   33
13. SKU-Order Completion Profile   35
14. Cubic Movement Profile   35
15. Shipment Variation Analysis   36
16. Product Selection Based on Historical Demand   37
17. Unit Load Characteristics and Their Impact on the Cross Dock Operator   38
18. Delivered Cost Comparison for Candidate SKU   41
19. Phase 1, Step 2b: Supplier Review   42
20. Form for Gathering Supplier Information   44
21. Supplier Evaluation   45
22. Summary of Supplier and Product Review   47
23. Phase 1, Step 2c: Audit of Current Operations and Policies   49
24. Phase 1, Step 3: Analyze Strengths and Weaknesses   52
25. Vision Checklist   53
26. Operations Review for Fictional Retailer   54
27. Facility and Equipment Review for Fictional Retailer   55
28. Information Systems Review for Fictional Retailer   56
29. Customer (Store) Review for Fictional Retailer   57
30. Transportation Review for Fictional Retailer   58
31. Comparing Cross Docking Systems   69
32. Automated Pallet Load Cross Docking System   70
33. Automatic Pallet Unloading Conveyor   71
34. Information Cycle for Pre-allocated Supplier Consolidation   73
35. Ideal Facility for Pure Supplier Consolidation (Full Pallet Movement)   74
36. Mechanized Cross Dock System Layout   77
37. Conveyor Equipment for Receiving and Sorting Cross Docked Products   78
38. Equipment for Sorting Cross Docked Products   79
39. Conveyor Equipment for Loading Cross Docked Products   79
40. Information Cycle for Pre-allocated CDO Consolidation   80
41. Product Flow for Pick-to-Belt with Automatic Sortation System   85
42. Information Cycle for Post-allocated CDO Consolidation   87
43. Suggested Flow Pattern for Combining Cross Dock with Storage   88
44. Conceptual Layout with Cross Docking and Pick from Inventory Modules   92
45. Phase 2, Step 1: Generate System Design   93
46. Cross Docking Systems and Their Operating Scenarios   94
47. Phase 2, Step 2: Perform an Economic Analysis on the Alternatives   96
48. Estimated Costs of Common Cross Docking Equipment   97
49. Equipment Costs of Operation and Ownership Worksheet   98
50. Phase 3, Step 1: Create the System-wide Cost Model and Determine ROI   107
51. Incremental Supplier Costs for Fictional Retail Chain   108
52. Annual Expenses by Category   109
53. Calculating Storage and Handling Expenses Per Carton   110
54. Costs for System-wide Model   111
55. Sample Model for Determining Cost Savings Across the Supply Chain   113

56. Start-up Costs for Cross Docking Initiative   115
57. SKU Cost Model Utilizing Direct Product Profit Method   117
58. Activity-Based Costing Concepts   118
59. ABC Model Example   119
60. SKU Comparative Cost Model/Activity-Based Costing Method   120
61. Phase 4: Implementation   124
62. Sample Project Implementation GANTT Chart for Cross Docking   127
63. Plans for Proposed Cross Docking System   128
64. Sample Process Flow Chart for Receiving   129
65. Sample Operational Plan Checklist   129
66. Sample Facility Plan Checklist   130
67. Sample Equipment Plan Checklist   130
68. Information Systems Plan Checklist   131
69. Transportation Plan Checklist   132
70. Product Specification Plan Checklist   133
71. Contingency Plan Checklist   133
72. Design Parameters for Cross Docked Items   149
73. Conceptual View of a Store Lane for Case Study   153
74. Existing Layout for Case Study   153
75. Alternative 1 Layout for Case Study   154
76. Alternative 2 Layout for Case Study   155
77. Labor Requirements for the Cross Dock of Full Cartons Only   156
78. Capital Costs of Equipment and Information Systems   158
79. Annual Operating Costs   159
80. Economic Analysis of Saw & Order   160
81. Product Flow Patterns   170
82. Sample Dock Layout   171

# Acknowledgments

This book could not have been produced without the valuable contribution of the following cross docking practitioners and experts:

- Mike Scott, Vice President, Distribution, Tops Markets, Inc.
- Virginia Carmon, IBM Global Services
- Jan Young, Director-Business Development, Catalyst International
- Mike Albert, Director of Operations-Consumer Sector, Exel

There are other logistic professionals who remain unnamed but whose contribution to this research project was just as important. The author wishes to thank all of these forward-thinking leaders in the industry for selflessly sharing their knowledge and experience on cross docking.

In addition, the following persons deserve special mention for their support.

**The WERC Task Force on Cross Docking:**

- Rita Coleman, for being the motivating force, for her editing support and encouragement
- Arnold Maltz, Ph.D., Assistant Professor, Supply Chain Management, Arizona State University, and Thomas W. Speh, Ph.D., Director, Warehousing Research Center, Miami University of Ohio, for reviewing pages and pages of manuscript, for sharing their logistics expertise and providing most constructive criticism and support

We also thank:

- Anura DeSilva, Ph.D., Planmatics, Inc., and his staff for their contribution to this work
- Kenneth Ackerman from the K. B. Ackerman Company, for his research support
- George and Roz Gross, for making it happen and for coming out of retirement to edit the manuscripts

- Don Derewecki, for his grocery "connections" and for his editing support
- Jack Kuchta, for his practical and successful approach to cross docking
- Geoff Sisko, for his logistics advice and support
- Robert Silverman, for his encouragement and for being the *askBob.com* of warehousing
- Chris Emrick, Tushar Patel, Eric Miller, Carlos Bastos, Stacey Ward, and Virginia Muller for their in-office support
- And for the rest of the staff of Gross & Associates who have each contributed in some way or another to this book

# Foreword

## A Tactical Approach to an Operational Strategy

Cross docking—almost as old as warehousing and brand new as we move into the 21st century. The truth of the matter is, though, that cross docking as a supply chain strategy has many strengths and possibilities. Like many other strategies, however, it's easier talked about than implemented.

This publication is a guide to tell you "how to do it."

*Making the Move to Cross Docking: A practical guide to planning, designing and implementing a cross dock operation*—the title says it all. In easy-to-understand terminology and detailed explanations, Maida Napolitano and the team at Gross & Associates have taken the complex topic and laid it all out for you. If you're just beginning to explore cross docking, you'll find a step-by-step guide to help you begin. Are you already doing some cross docking? Great! Use this reference to fine-tune or expand your operations. Are you teaching cross docking concepts or trying to sell the possibility to management? This text provides the information you need.

WERC takes pride in providing practical, down-to-earth information for its members, and this *Guide* is no exception. Distribution and warehousing professionals are realists and this publication speaks to that mindset. It is solidly research-based and describes many different applications and real-life experiences. It does not hold cross docking as some kind of panacea, but candidly points out possible drawbacks and difficulties as well as strengths and opportunities. As the text points out: "Cross docking cuts across many interacting functions within, and beyond, the four walls of the warehouse. It takes on a wide variety of forms—from simple full pallet movement to complex sortation systems. Determining the most appropriate operation can be a quagmire of details and data."

We are pleased to be able to offer this definitive text on such an important warehousing operational strategy.

Bill Miller
Manager, Marketing and Sales
Gateway Warehouse Company, Inc.

# How to Use This Book

This book was written with practicality in mind. Its goal is to provide novices and experts with a guide for making the transition to cross docking. A conscious effort was made to provide specific goals at each phase in the process, which can be easily accomplished, and not to inundate readers with logistics philosophies or mathematical models.

**Chapter 1** provides an overview of cross docking and how it evolved into a value-added logistics strategy. It expands on the strategy's benefits and drawbacks and describes numerous applications, including a discussion of its relevance to current trends in the industry. **Chapter 2** outlines the steps for *assessing* if one's business is capable of sustaining a formal cross docking program. It includes an in-depth look into product profiles, preliminary costs, supplier performance, and a business' current infrastructure so that a determination can be made on whether cross docking would be effective. **Chapter 3** describes different types of cross docking systems and discusses the steps for *designing* a system that will best suit a company's needs. **Chapter 4** provides guidelines for *cost justifying* cross docking to management. However, if the math has been done and cross docking *is* determined to be the answer, then the reader may simply move on to **Chapter 5**, which details the steps for *implementing* cross docking. Chapter 5 also provides a summary of pitfalls to avoid when designing and implementing cross docking. **Chapter 6** presents a case study of a real life retailer's initial foray into cross docking—from design to implementation.

For illustrative purposes, actual cross docking success stories, caselets, and vignettes from interviews and from various print media have been included throughout the book.

> Chapters 2 through 5 detail the four phases to make the transition into cross docking. A reader who is seriously considering adopting this strategy should focus on these chapters.

The Appendices provide:
- A list of resources (both print and Web-based)
- Because cross docking requires well-designed docks, a narrative on general dock design

# CHAPTER 1

# Cross Docking at a Glance

**Objectives of Chapter 1:** *To provide an overview of cross docking. To emphasize the benefits (especially cost savings) of cross docking. To include a guide on how to use this book whether you are a manufacturer, wholesaler, third party provider, retailer, etc.*

**1.1** Defining cross docking

**1.2** A brief history of cross docking

**1.3** Types of cross docking applications

**1.4** Ideal conditions for success in a cross docking program

**1.5** Cross docking cannot exist alone

**1.6** Choosing between cross docking and other distribution strategies

**1.7** Benefits and drawbacks of cross docking

**1.8** Applying cross docking to your operations

**1.9** Trends that stress its importance in a changing logistics industry

**1.10** Summary

# CHAPTER 1

# Cross Docking at a Glance

Welcome to the new millennium—an age where businesses are tightening their belts, increasing their pace, and working closely with radical advances in information technology. Customers are demanding speed and accuracy, and the warehouse is undergoing a transformation—from a stolid physical facility used to house "mistakes" in forecasting called *inventory* to one that distributes products in the shortest possible time at the lowest possible cost. Products will flow through this new warehouse with hardly any storage and rarely any picking. This could be the end of warehousing as we know it. Welcome to the strategy of *cross docking*.

The biggest irony may be that cross docking has been around for decades. Why is there a resurgence in interest? Why are more companies turning to this relatively "old" strategy to face the future? There are a number of reasons.

## Accelerated Product Flow

Cross docking speeds the flow of product through the supply chain. Products are routed toward their end destination as soon as they are received. They do not sit in a warehouse as inventory waiting for orders.

**REAL WORLD EXAMPLE**

### Hannaford Feels the Need for Speed in Perishables

Hannaford Bros., a Scarsborough, ME-based distributor of grocery items reports great potential in the cross docking of perishables such as pork, chicken, and some red meat. The short shelf lives of these items require that they be handled promptly, making them good candidates for the program. These items are all high-volume, high-velocity products that are routinely purchased by stores in fairly large quantities. Hannaford receives perishables as close to the outbound scheduling window as possible and immediately picks store orders off the pallets at the time of receiving. These orders are sorted to store lanes, and as pallets of outbound product are accumulated, they are loaded onto the trucks.

Another perishable that is cross docked on a daily basis is seafood. Seafood items come off the boat, across Hannaford's dock, and are almost immediately loaded for delivery to stores. With cross docking, freshness is guaranteed. (Source: Casper, Carol, "Flow-Through: Mirage or Reality," *Food Logistics*, October/November 1997, p. 50.)

CHAPTER 1   *Cross Docking at a Glance*

## Cost Savings

Cross docking offers significant reductions in bottomline costs. With product received and shipped immediately, storage is eliminated and inventory is reduced. Consequently all the costs associated with handling and holding that inventory are also reduced or eliminated.

**REAL WORLD EXAMPLE**

### Inventory Reduction at Fred Meyer

Fred Meyer is a full-line discount department store. It looked at reducing costs in its distribution center (DC) by organizing suppliers to support a cross docking strategy. Working with 30 vendors of 609 SKUs, the company developed standards for cross docking that took 1.4 million inventory dollars out of the DC. In addition, seven suppliers of 1,322 SKUs had the capability of consolidating product by store, thus taking $1.09 million more out of its DC inventory. When furniture arrived in bulk from yet another supplier, the company allocated it store-by-store as soon as it was received at the dock. As this supplier became capable of packing by store, Fred Meyer saved $200,000 in DC labor costs in addition to saving $500,000 in inventory by cross docking the supplier's products. With cosmetics, the company saved an additional $350,000 in DC inventory costs by eliminating 450 SKUs after a vendor developed a pick-and-pack-by-store flow-through program. Using vendor-managed inventory with two key food suppliers, inventories declined 30 to 40 percent while service levels increased to 98 percent. Because of just-in-time delivery and cross dock of SKUs, warehouse personnel put product directly into the pick slot rather than storage, thus reducing DC labor costs. Its new flow-through distribution center has room to store only 1,500 of the 6,500 SKUs that flow through it. Instead of taking three days to ship after receiving, the new center ships within 24 hours. Most orders are pre-allocated to specific stores and simply cross docked through its facility. Where cost-effective, Fred Meyer uses third party providers and frequently sits down with suppliers to work out programs that are mutually profitable for both parties. (Source: Halverson, Richard, "Fred Meyer Defines QR Success," *Discount Store News*, April 17, 1995, p. 6.)

## Faster, Cheaper Routing

Cross docking provides a faster and less expensive way of routing shipments from suppliers to customers by consolidating products bound for the same destination. In many cases, less expensive and faster modes of transportation can be used.

**REAL WORLD EXAMPLE**

### Ford Automotive's Mixing Centers

Most new automobiles manufactured in the United States are transported by rail from manufacturing plants to special railroad centers, called ramps, and then by truck to local dealers. At the plants, newly assembled automobiles are parked in load lanes according to a destination ramp. When a specific number of vehicles have accumulated, they are loaded on a tri-level railcar, capable of carrying 15 sedans, 5 on each deck. In the past, these railcars passed through switching yards, where those headed in the same direction were sorted onto trains. In crossing the country, a railcar may pass through half a dozen switching yards before reaching its final destination ramp. At the destination ramp, vehicles are unloaded and parked to await delivery to their designated dealerships. When enough cars and trucks for dealerships in a given area have accumulated, they are loaded onto a transport truck and delivered to dealers.

In recent years, Ford implemented a new network involving four special cross docking centers across the country, called mixing centers, in partnership with Norfolk Southern. By 1998, all 21 of Ford's North American assembly plants were using these centers, which are expected to handle 3 million vehicles annually via a network of 19 new unit trains. Unit trains consist of 20 or more railcars with a common destination, bypassing switching yards. In each mixing center, vehicles on railcars from different assembly plants are unloaded, sorted by destination, and loaded back onto railcars with other models bound for the same destination. Previously it took an average of 12 days to deliver vehicles from the plants to the dealerships. This new mixing center network was designed to cut this transit time by one-third. Routing shipments through a mixing center can reduce transportation time and cost in two ways: *(1) Allow faster transport.* Consolidating shipments at a mixing center generates sufficient volume to warrant the use of the faster unit trains, which bypass switchyards. (2) *Reduce wait.* Because the daily supply rates to some ramps are much smaller than the capacity of a railcar, automobiles destined for these ramps may have to wait several days at an assembly plant's load lane for a full load. Consolidating shipments from multiple plants to a mixing center can increase volume and reduce delays. Eliminating load lanes at the plants also frees up valuable pace for more productive uses. Each plant will now be making room for 5 mixing centers instead of the previous 15 ramps. By early March 1999, the system was nearly running up to speed, on budget, with the technology in place. (Sources: Ratliff, H. Donald, et al., "Network Design for Load-Driven Cross Docking Systems," Research paper from the Georgia Institute of Technology, pp. 1-2, and on the Web under Research at http://www.tli.gatech.edu/lms2000/; and Stephens, Bill, "Roads smooth out for NS auto hubs," *Trains*, June 1998, pp. 21-22.)

## Supports Customer Needs

Cross docking allows firms to meet and support specific needs of customers such as just-in-time practices, the consolidation of multiple supplier networks, the launching of specific promotions and other marketing strategies.

**REAL WORLD EXAMPLE**

### Cross Docking Supports Just-In-Time for Mitsubishi and GATX Logistics

GATX Logistics and Mitsubishi Motor Manufacturing of America have been working together for over a decade. Mitsubishi wanted to create a more even flow of components to the assembly line, promote a safer work environment, and maximize the space within the assembly plant. Growing production volumes and the changing mix of cars led to insufficient on-site storage and a strong need to coordinate the supply chain process. To solve these problems, GATX designed and implemented a cross dock operation in Normal, IL, adjacent to the Mitsubishi assembly plant. The cross dock center handles the flow of 8,500 production components and service parts inbound from 350 of Mitsubishi's nearly 400 vendors. Two production shifts at the plant work a combined total of 16 hours per day, five days per week. The necessary component parts—approximately 3,600 daily parts orders—must arrive for production shifts in prompt two-hour intervals, no less than two hours prior to scheduled assembly. At the cross dock center, GATX personnel may sort, pre-assemble, and sequence components from multiple vendors. Then they stage consolidated components for delivery into the assembly plant just when they are needed. A one-day safety margin—or buffer stock—is built into the handling process. (Source: Witt, Clyde E., "Crossdocking: Concepts Demand Choice," *Material Handling Engineering*, July 1998, p. 48.)

## A Fundamental Shift

Despite these benefits, not many companies are cross docking. Although simple in concept, cross docking is a fundamental shift in thinking; and for some warehouse managers, the idea of not storing product in anticipation of demand is difficult to grasp. Another major obstacle is not knowing how to make it happen. Cross docking cuts across many interacting functions within, and beyond, the four walls of the warehouse. It takes on a wide variety of forms—from simple full pallet movement to complex sortation systems. Determining the most appropriate operation can be a quagmire of details and data. The key to success is adopting a systematic approach in planning, design, capital justification, and implementation. This publication showcases one such approach.

This chapter defines cross docking, expands on its strengths, discusses its evolution in the industry, and argues its relevance to current logistics and business trends. The remainder of the book will focus on each of the steps for making the actual transition into cross docking in four successive phases:

**Phase 1:** Assessment and Negotiation
**Phase 2:** Planning and Design
**Phase 3:** Justification and Cost Sharing
**Phase 4:** Implementation and Maintenance

Setting up a cross docking program can be an enormous challenge. But in today's fast-paced business world, the consequences of not considering it can be detrimental. Three- or four-day order cycle times are a thing of the past. Supply chains are looking for accelerated flows. An order received by a supplier as late as 2 p.m. should be picked in a distribution center by early evening, shipped to the retailer's DC by 4 a.m., cross docked, and shipped to the store by

> THE KEY TO SUCCESS IS ADOPTING A SYSTEMATIC APPROACH IN PLANNING, DESIGN, CAPITAL JUSTIFICATION, AND IMPLEMENTATION.

2 p.m. the next day. Speed is the name of the game, and cross docking is the tool to make it happen. It may not be a new idea, but it certainly deserves a closer look.

## 1.1 Defining cross docking

Defining cross docking may not be as easy as it seems. Research shows that different people in the industry have come up with their own definitions for the term. Some define it as *the movement of full pallets from inbound trucks directly to outbound trucks with the load never touching the warehouse floor.* Many industry analysts think that this definition is too limiting. Not everyone can afford the support system for this method of cross docking. Then there is the all-encompassing definition: *Cross docking is a collection of value-added activities that occurs all along the supply chain.* That makes cross docking just about anything.

For this publication, a definition is based on analysis of printed material and interviews with cross docking practitioners and experts in the field from various industries. Cross docking is defined *as a process where product is received in a facility, occasionally married with other products going to the same destination, then shipped at the earliest opportunity, without going into long-term storage. It requires advance knowledge of the inbound product, its destination, and a system for routing the product to the proper outbound vehicle.* See Figure 1.

Three distinct characteristics have to be evident before an operation can be considered cross docking.

1. The time that merchandise is staged in either the receiving or shipping dock *must be kept to a minimum.* Some experts argue

**FIGURE 1**

**Defining Cross Docking within a Facility**

Cross docking is defined as:
A process where product is received in a facility, occasionally married with other products going to the same destination, then shipped at the earliest opportunity, without going into long-term storage. It requires advance knowledge of the inbound product, its destination, and a system for routing the product to the proper outbound vehicle.

- No storage
- No delay
- System in place for information exchange and product movement

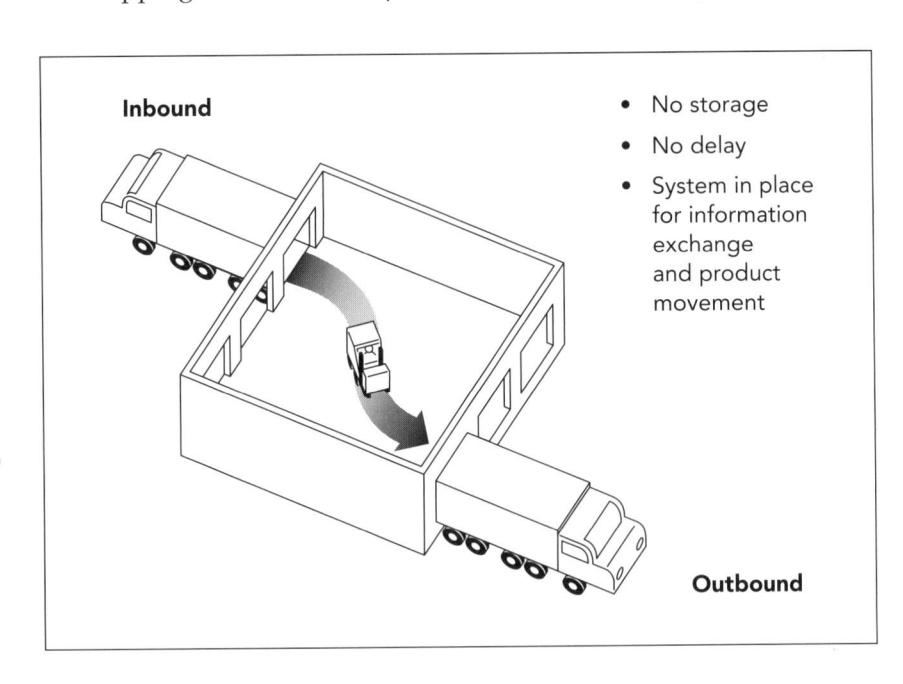

that if product sits in the staging area for more than a day, then the process should not be considered cross docking. However, some third party services do not start charging storage fees unless the merchandise is still in the facility after three days.[1] In addition, some delay between shipping and receiving may be inevitable, as product may need to be staged to suit a store's once-a-week delivery schedule. Some merchandise may have to be held in trailers at the dock waiting to be cubed out. The bottom line is that every effort should be made to facilitate the flow of product from the receiving trucks to the shipping trucks in a minimum amount of time.

2. After receipt, the product must either go directly to shipping or stay on the dock or forward picking area and *never be put into reserve storage*. With cross docking, storage is essentially eliminated as product is immediately processed for shipment before or upon its receipt.

3. *A system must be in place* to ensure the efficient and effective exchange of physical product and information. There should be coordination among the different cross docking players with considerable attention given to the matching and scheduling of inbound receipts to outbound shipments so that product does not stay in the facility longer than it has to. The ideal cross dock system should be able to process incoming merchandise and route its next destination in the supply chain—even before receipt of the physical product.

### FIGURE 2

**Generic Definition of Cross Dock Players**

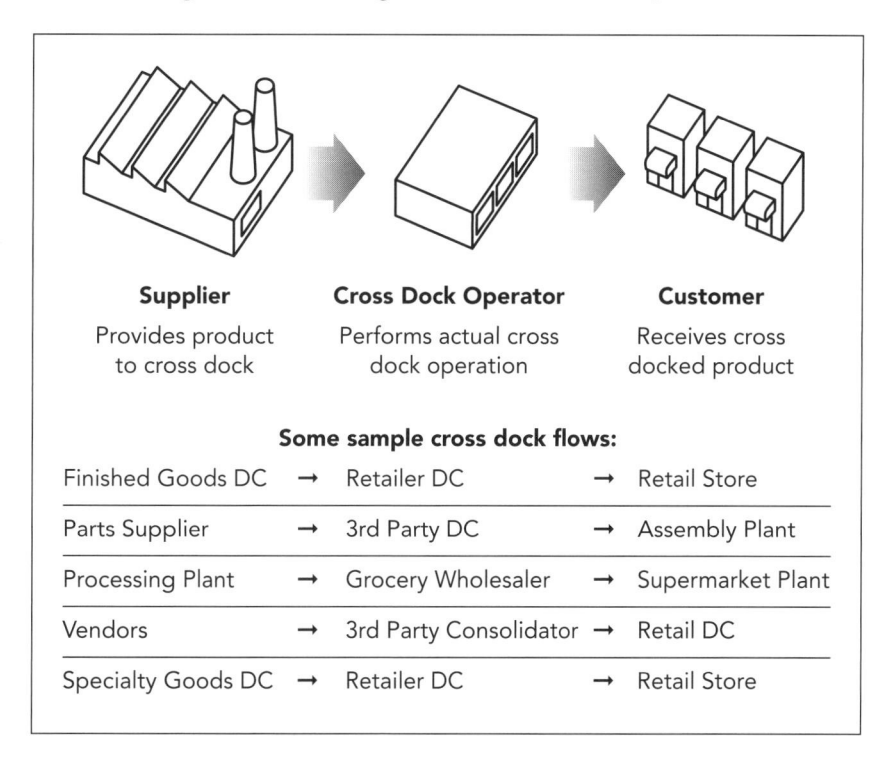

| Supplier | Cross Dock Operator | Customer |
|---|---|---|
| Provides product to cross dock | Performs actual cross dock operation | Receives cross docked product |

**Some sample cross dock flows:**

| | | | | |
|---|---|---|---|---|
| Finished Goods DC | → | Retailer DC | → | Retail Store |
| Parts Supplier | → | 3rd Party DC | → | Assembly Plant |
| Processing Plant | → | Grocery Wholesaler | → | Supermarket Plant |
| Vendors | → | 3rd Party Consolidator | → | Retail DC |
| Specialty Goods DC | → | Retailer DC | → | Retail Store |

## Defining Terms

There are many applications of cross docking that span different trading points in the supply chain. Throughout this text, generic terms are used for the main players involved. The term *cross dock operator* or *CDO* refers to the party that performs the actual cross docking within his facility. A CDO may be a third party provider moving manufacturing components from multiple vendors across the dock to assembly plants, or it may be a retailer cross docking product to its own stores. It could also be a manufacturer, with multiple production plants—each making a different product. The manufacturer marries up the various items in a cross dock operation without putting any of them into storage.

The term *supplier* will refer to the party providing the product to be cross docked. A supplier may be a vendor moving the product to a retailer's distribution center, or a third party provider assembling multi-SKU pallets for cross docking at a retailer's distribution center.

The term *customer* refers to the party who will receive the cross docked merchandise at the final destination. Items may be received in a retail store for sale to the consuming public or in a manufacturing plant for use in assembly and production. See Figure 2.

## 1.2   A brief history of cross docking

The U.S. Postal Service (USPS) was perhaps one of the earliest practitioners of cross docking in the United States. Although the Postal Service started out primarily delivering letters, the onset of mail order business in the late 1800s saw a tremendous increase in package distribution. Montgomery Ward started its first catalogue business in 1872, and Sears, Roebuck and Company quickly followed suit, boasting that it was selling four suits and a watch every minute, a revolver every two minutes, and a buggy every 10 minutes—all of these items shipped by slow, but reliable, parcel post. In those days, the packages were sent from the catalogue companies through the official post office relay stations where they were cross docked and delivered via railroad to catalogue customers. Today these relay stations have become state-of-the-art sortation facilities cross docking millions of packages per day. With cartons going in and out the same day, USPS and other package service companies such as United Parcel Service and Federal Express now represent the cutting edge of cross docking applications.

In the distribution arena, early cross docking was called "expediting." Expediting was done to satisfy some critical need or cater to demands of a desperate customer, carried out on an ad hoc basis with a few packages here and there.[2] With the growth of the railroad industry, businesses applied cross docking in the transport of their pool car shipments to distribution facilities, commonly called freight houses. A pool car was a rail car loaded with products for several customers. Freight houses had rail siding on one side, a truck dock on another.[3] Warehouse workers would unload this pool car and move the product *across the dock* to the trucks on the other side of the warehouse. Products from a pool car might be combined with other items bound for that customer, before loading on a truck to be sent

CHAPTER 1   *Cross Docking at a Glance*

off to the customer's warehouse. A number of major industries still use this pool car concept today.

The growing popularity of Just-In-Time (JIT) in manufacturing placed cross docking in the forefront of change. With JIT, components have to be made available at a specific location just when they are needed for a production process to continue. Managers realized that instead of keeping several days' supply of parts on hand for assembly, the inventory buffer could be reduced to hours. Logistics managers quickly saw how the JIT philosophy could be applied to the distribution facility by receiving shipments just when they are needed for shipment. This translated to dramatic reductions in inventory. Simply put, JIT in the distribution arena became *cross docking*. It was quickly transformed from a method of facilitating the movement of a few packages to a full-fledged logistics operation. By the early 1980s, mass merchandisers, such as WalMart and Kmart, formally adopted cross docking as part of their logistics operations. By 1982, Kmart was cross docking soft goods and in 1987 added hard goods. Since 1994, about 90 percent of soft goods are being cross docked through Kmart's distribution centers.[4]

## 1.3   Types of cross docking applications

A variety of cross docking applications is routinely undertaken at different points in the supply chain.

### Manufacturing Cross Docking

In the manufacturing arena, a manufacturer or a third party provider working for the manufacturer may use cross docking for the receipt, consolidation and shipment of a pre-defined quantity of raw materials or component parts, typically from many suppliers, to the production plant. Only the required quantities of parts for a pre-scheduled production run are shipped hours before they are needed on the assembly line. Because of the rapid response nature of JIT, suppliers and cross docking centers are usually located within miles of the manufacturing assembly plant. See Figure 3.

**REAL WORLD EXAMPLE**

#### Miller SQA, Menlo Logistics, and JIT

Miller SQA (Simple, Quick, and Affordable), Inc., a furniture manufacturing subsidiary of Herman-Miller Inc., produces a wide variety of low-cost, quality office products on a build-to-order basis. It has become a leader in furniture manufacturing with a lead time average of seven days, compared to the industry standard of three to six weeks. With hundreds of thousands of product variations and the demands of a build-to-order system, scheduling component parts is not easy. To protect stock-outs and production delays, 15 days' worth of parts inventory was previously kept on its shop floor. This increased SQA's carrying costs and used up 40 percent of the plant's floor space. In addition, SQA's business was expanding at 30 percent a year, and space was desperately needed to increase production capacity.

To improve their JIT processes, the company decided to bring the supply base closer and hired Menlo Logistics to manage a single parts warehouse just four miles from the Holland production facility. Suppliers ship more than 1,000 different parts to the site, where Menlo employees pick, sort, and deliver parts to the Holland plant based on SQA'a actual production needs. This cross dock center is notified of the plant's needs four hours in advance. Menlo uses two hours to pick the parts, load them into discrete containers, and deliver them to the plant. The parts are either shuttled directly to one of 16 work centers on SQA's shop floor or placed in a staging area until needed. At any given time, there is no more than two hours' of production part inventory at the SQA plant. This frees up valuable space for new production lines and helped SQA reduce significant inventory carrying costs.

In addition, suppliers, not SQA, own the inventory at Menlo. But it has not simply shifted its inventory downstream. In fact the opposite is true. Suppliers are required to keep no more than 10 days worth of inventory in their facilities. In addition parts do not stay in the cross dock center more than two days. They are geared to simply flow through Menlo's facility.

SQA admits that the flow of information from customers to suppliers is the key to the success of their JIT program. Once they know when an order needs to ship, a manufacturing software program projects the time

**FIGURE 3**

**How Cross Docking Supports Just-In-Time (JIT)**

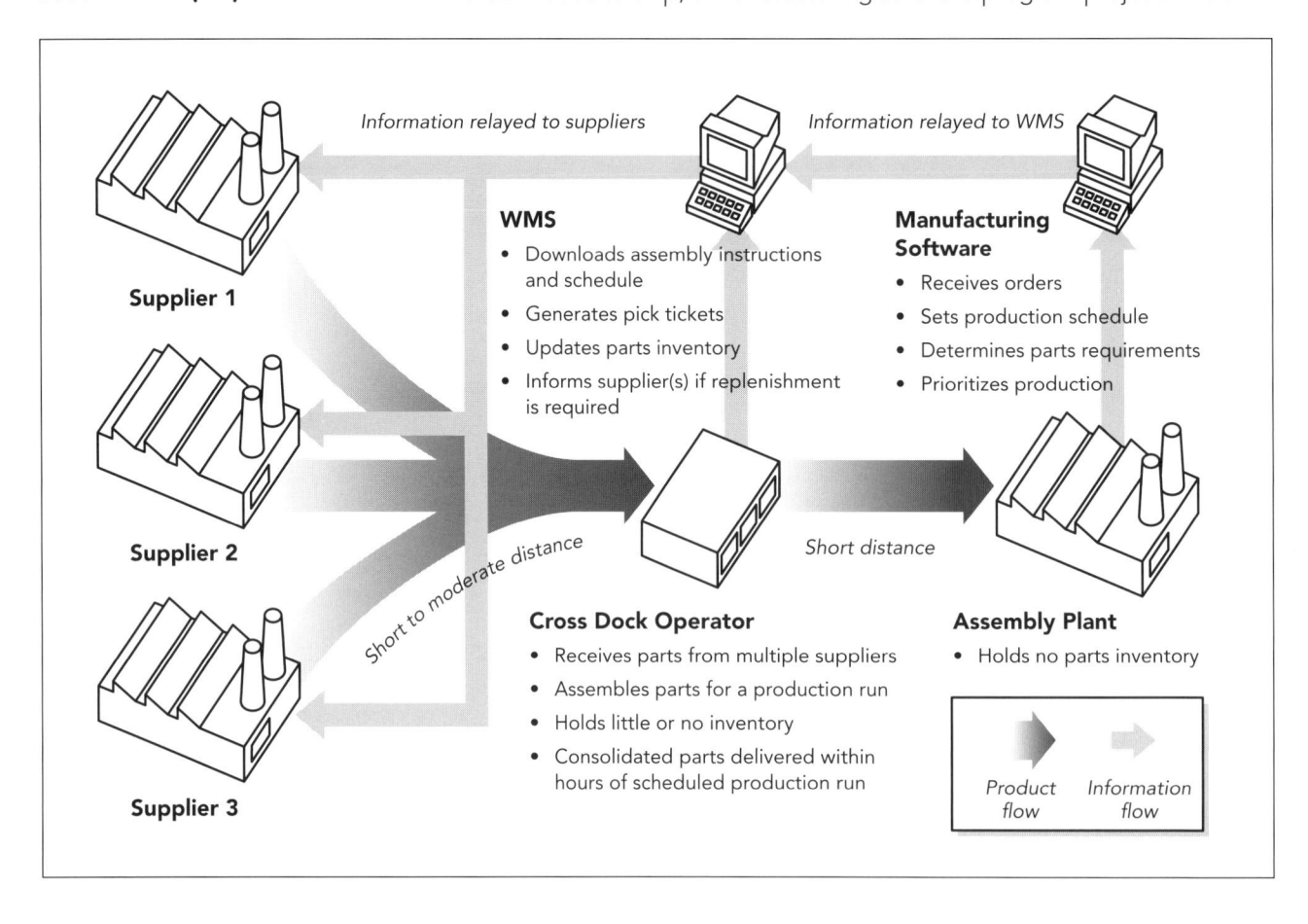

CHAPTER 1   *Cross Docking at a Glance*

necessary to complete an order and prioritizes production accordingly. When the production order is set, the system electronically transmits part demands, production queues, and delivery schedules to Menlo's warehouse management system. Barcode labels attached to the discrete bins tracks the type and number of parts in the tote. SQA assembly line workers scan the barcode when the tote is empty. The software uses this scanning information to determine when to ask for more parts from Menlo and when to place new orders with the suppliers. (Source: Minahan, Tim, "Miller SQA tweaks JIT system for quick response," *Purchasing*, September 4, 1997, p. 48.)

**FIGURE 4**

**Post-Production Cross Docking**

Adapted from: Moore, Thomas and Roy, Chris, "Manage Inventory in a Real-Time Environment," *Transportation & Distribution*, July 1998, p. 70.

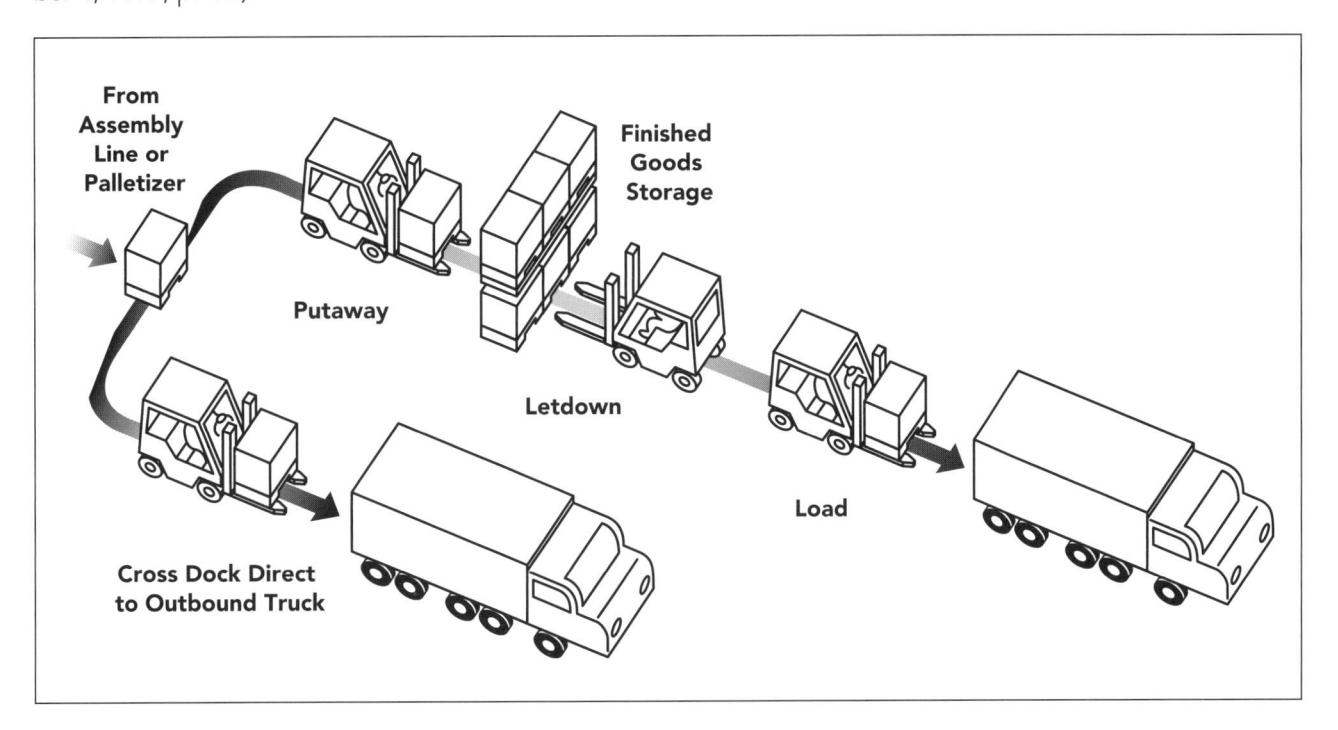

Although less complex than other cross docking applications, the process of directly loading finished goods into outbound trucks *after* the production process is also considered by some people as cross docking.[5] To save valuable space and handling costs, manufacturers are not putting product into a finished goods warehouse immediately after production; instead they are cross docked directly to outbound trucks, which transport the product to the next level of the supply chain. See Figure 4.

REAL WORLD EXAMPLE

### Even Beer Gets Cross Docked After Production

Cross docking is a key practice at Miller Brewing in Milwaukee, WI. In each brewery, most cases of beer move from a palletizer, where they are ticketed, directly to trailers by forklift. A McHugh Warehouse Management System, LXE (www.lxe.com) radio frequency device and SYMBOL long distance scanners track an average of 11,000 pallet moves a day at each of Miller's six breweries from the palletizer, to quick staging positions on the floor (when needed), then to trailers on their way to distributorships. Miller is down to an average of two handlings per pallet. In the past, everything went to the floor and was handled several times. (Source: Lear-Olimpi, Michael, "Looking for the fast lane," *Warehousing Management*, April 1999. p. 26.)

## Distributor Cross Docking

At the distributor's facility, cross docking is performed on a wide variety of merchandise. Typically it is employed in operations where various manufacturers, making complementary items, ship their merchandise to a common distributor who assembles the products on a multi-SKU pallet before delivery to the next level of the supply chain.

> **REAL WORLD EXAMPLE**
>
> ### Assembling Product From Different Publishers at Rand McNally Logistics Solutions
>
> Rand McNally Logistics Solutions performs cross docking for Rand McNally's publishing operation, for publishers who have printed material produced by Rand McNally, as well as for companies in other industries. Logistics Solutions and its third party provider receive full pallet quantities of books from these different publishers, marry titles, package and label each shipment, then sort it to any one of U.S. Postal Service's 21 bulk mail centers, a center for United Parcel Service, or a priority shipper. The shipments are loaded in truckload quantities at Logistics Solutions' hub-and-spoke center in Lexington, KY, and then delivered to the appropriate bulk mail center. The Postal Service or UPS then delivers the books to individual addressees. (Source: Harps, Leslie Hansen, "Crossdocking for Savings," *Inbound Logistics*, May 1996, p. 33.)

## Transportation Cross Docking

Although it is not the main thrust of this publication, a discussion on cross docking would not be complete without mentioning the cross docking services provided by transportation providers such as UPS, Federal Express and other regional carriers. They sort and consolidate parcels and pallet loads based on geographic destination. Consolidation typically reduces transportation costs by accumulating full truckloads for a specific region in a facility. Delays, attributed to waiting for sufficient volume to accumulate, are also reduced by converging merchandise from multiple sources. Most of these operations are fairly manual and typically cross dock full pallet quantities. Parcel delivery companies, however, are more automated with each package pre-labeled with its final destination and pre-weighed to facilitate its entry into the system. In both cases, all of the merchandise received is sorted by geographic area, and shipped in a matter of minutes to hours.

> **REAL WORLD EXAMPLE**
>
> ### Cross Docking at UPS
>
> The United Parcel Service (www.ups.com) facility at the Philadelphia International Airport is just over one million square feet with 675,000 square feet of staging space and 170,000 square feet of mezzanine. It is the parcel delivery company's second largest air hub and third largest facility. On average 300,000 packages flow through the DC from and into trucks and aircraft containers. These packages move with the support of 1,500

conveyors covering 10 miles and 3,130 workers. Nine input conveyors receive approximately 44,000 packages from air containers each hour. Seventy-seven people sort 31,000 parcels from 33 trailers at a rate of 1,000 packages an hour into two primary slides. The operation is fairly manual, but the company is looking to automate the movement of small packages of six pounds and less, with dimensions of 11 by 14 inches. These parcels will be bar coded, pass underneath an overhead scanner, be directed into a particular slide, and go directly into outbound bags, ready to be tagged and sent out. This scanner will eliminate current manual radio-frequency gun scanning on inbound packages. For outbound processing, an overhead scanner is already in place to verify that packages are going to the correct container. (Source: Lear-Olimpi, Michael, "Looking for the fast lane," *Warehousing Management*, April 1999, p. 28.)

Transportation providers may also provide the assembly process (sometimes carried out by the manufacturer or distributor) as part of their transportation service. An increasing number of manufacturers are using these companies and their facilities in an evolved form of cross docking, called *merge-in-transit*. An example of a merge-in-transit application can be seen in a partnership between Dell Computer Corp. and UPS. In Dell's plants, employees configure custom software and hardware on a computer, based on a customer's specific needs, while the monitor is manufactured and stored in another facility. At the same time that Dell ships the computer to a UPS facility, the monitor is shipped from another facility to the same UPS facility. The monitor and computer are eventually cross docked and merged in transit within the UPS delivery system.[6]

## Retail Cross Docking

Supermarkets, department stores, mass merchandisers, and warehouse clubs (such as Sam's Club, BJ's, and Costco) are a few examples of companies in the retail industry that do a tremendous amount of cross docking. Products are received at these retailers' distribution centers, moved across the dock, and married with other products bound for the same store. In some cases, suppliers provide display-ready merchandise and pallet assortments so that the retailer simply moves full pallets from receiving docks to shipping docks in a matter of minutes. Another form of cross docking in a retailer's DC may involve the placement of high-volume and/or seasonal merchandise immediately upon its receipt at a "hot processing area" where cartons for store orders are picked, labeled, then routed through an automated sortation system, where they are palletized with other products with the same store destination.

**REAL WORLD EXAMPLE**

### Cross Docking at Meijer

Meijer is a Grand Rapids, MI, operator of 100 superstores that use cross docking with more than 50 grocery vendors and more than 500 general merchandise suppliers. Cross docking candidates include promotional goods, seasonal items, and store-specific pallets, but no temperature sensitive products. All pallets and cases earmarked for cross docking must be received from manufacturers with UCC128 bar code labels. There is no manual processing of store orders at their cross dock centers. (Source: Garry, Michael, "Meijer's Caveats," *Progressive Grocer*, May 1996, p. 16.)

**Opportunistic Cross Docking**

The previous applications describe scenarios where cross docking is used as *standard business practice* or as part of a continuing operation to achieve a company's strategic goal of improving customer service at the lowest costs. This is in contrast to another popular application of cross docking called *opportunistic cross docking,* which is cross docking on an "as needed" basis. Manufacturers and distributors often use opportunistic cross docking to fill back orders when the product is received. *There is no pre-selected group of products for a specific period of time that will be cross docked.* In more sophisticated operations, the manufacturer may compare customer orders placed 24 to 48 hours in advance. Ship-to date merchandise is checked against scheduled receipts coming from production plants. Specific pallet quantities are identified to fill orders within the 48-hour period. Upon receipt, these pallets are cross docked through the manufacturer's facility to the next point in the supply chain. In the traditional warehouse, a WMS may provide decision-support capability to anticipate receipts and orders and recommend cross docking opportunities. In other cases, personnel can enter the system and identify pallets in the WMS to be cross docked in addition to a suggested cross docking report. Although these opportunistic methods of cross docking can save operational costs in the short term, they may *not* produce the level of benefits and savings of a formal cross docking program. The pre-selection of products and suppliers that can sustain long-term formal cross docking programs is the main thrust of this publication.

## 1.4 Ideal conditions for success

A formal program of cross docking cuts across many functions that must work together for the program to succeed. Using the generic definitions that were established for the different cross docking players, the ideal cross docking operation has the following essential elements (Figure 5).

**The Right Products**

An ideal cross docking program moves the appropriate products at reduced costs. These are products that are in pre-mixed or pre-consolidated pallets that require minimum handling upon receipt at

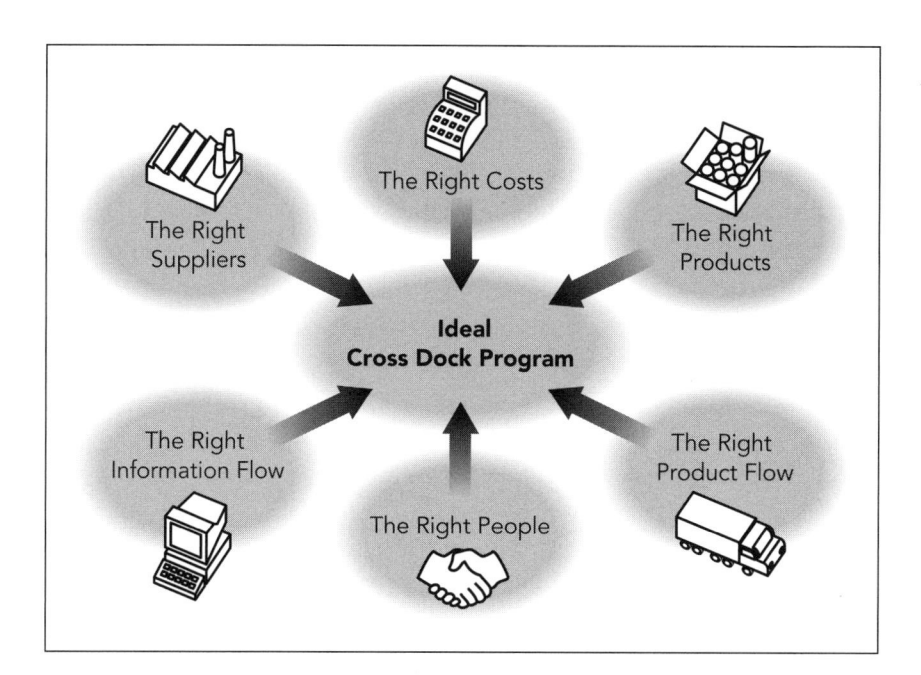

**FIGURE 5**

**Elements for the Ideal Cross Dock Program**

the cross dock operator's facility. These are products that have high inventory carrying costs, which can be significantly reduced with cross docking. These are products that have bar coded labels attached to the cartons or pallets with pertinent product information readily available so they can be routed immediately after a single scan. These are products that have predictable demand patterns so that their shipment can be accurately anticipated and properly scheduled.

## The Right Suppliers

Ideal cross docking should have suppliers who have the right processes in place so that they can consistently provide the correct quantity of the correct product at the precise time when it will be needed. These are the suppliers who are capable of configuring products for efficient handling through the next point in the chain, are always be in communication with the CDO, and are always providing necessary feedback. This capability of suppliers often determines the complexity of a cross dock operation for the CDO. If a supplier is able to label and pre-sort products on pallets, the CDO will be able to work quickly and ship with minimal handling. If not, sortation may be required, increasing the complexity of the operation.

## The Right Information Flow

Ideal cross docking must have a timely, accurate, preferably paperless information flow among trading partners and a smooth, continuous product flow that is matched to actual demand.[7] This information is used by the logistics planner to schedule the receipt of products at the CDO to coincide with the outbound shipments—shipments in response to the *demand* generated by the next customer in the chain. For raw material suppliers providing component parts just-in-time to production plants, demand would be actual production runs of the manufacturer. For the manufacturer supplying finished goods to a distributor, demand would be the actual orders by the distributor's retailers. For the distributor supplying bulk products to the retailer, demand would be aggregated store orders from a particular facility. The retailer's store orders should be based on actual consumer demand based on point-of-sale (POS) information. Accurate information and real-time inbound and outbound scheduling makes cross docking easier and more flexible.

**The Right Product Flow**

Ideal cross docking should have the appropriate network of transportation, facilities, equipment, and operations in place to support the flow of the product from the supplier through the CDO's facility to the customer. Where sortation is required, suitable material handling equipment and operations should be appropriately designed. Facilities should be configured to accommodate the operation— whether it is pure cross docking or a mixture of cross docking and traditional "store-and-ship" warehousing. Transportation should be set up to offer prompt and efficient support to operations in the facilities of each trading partner.

**The Right Cost**

As with any operating strategy, cross docking should only be performed if proper analysis and cost justification has been conducted. Any capital laid out to create the system should provide management with an acceptable and realistic return on investment.

**The Right People**

Ideal cross docking is carried out by personnel who recognize the urgency of moving product rather than storing it. They consistently track performance to determine strengths and weaknesses, and they acquaint themselves with the latest technology to determine how to best use it for their business. These are also people who are able to negotiate and coordinate with suppliers and customers.

## 1.5 Cross docking cannot exist alone

Cross docking is a distribution tactic to enable the fast and efficient movement of goods, but if inaccurate information or bad planning data is used, it cannot be effective. A company may install the most efficient cross docking system in the world, and yet its stores may still experience frequent out-of-stock or overstock conditions. How is that possible when products go in and out of the facility on schedule and without delays? Simple. Inventory management, transportation management,

CHAPTER 1   *Cross Docking at a Glance*

category management, procurement, and other peripheral functions may not have determined correctly *what, when,* and *how much* product to move. An example will better illustrate this critical point.

In a grocery retail operation, the logic for replenishing a premium brand of ice cream in a retailer's stores may be based on the fact that 40 percent of the U.S. population prefers the premium brand. But in stores that are situated in areas with household incomes over $100,000, 80 percent of the households may prefer the premium brand. These stores will have many frustrated—albeit rich—customers who may find zero stocks of their favorite ice cream and who will go elsewhere to shop. Clearly, the most efficient cross docking will be ineffective in improving customer service levels for this retailer if the wrong quantity or the wrong merchandise is shipped to the store's shelves.

## 1.6   Choosing distribution strategies

This research identifies cross docking as one of three possible options for moving product between trading partners in the supply chain. These three methods include (Figure 6):

1. **Traditional Method:** Product is received, stored, picked, and shipped from a facility.

**FIGURE 6**

**Different Distribution Strategies**

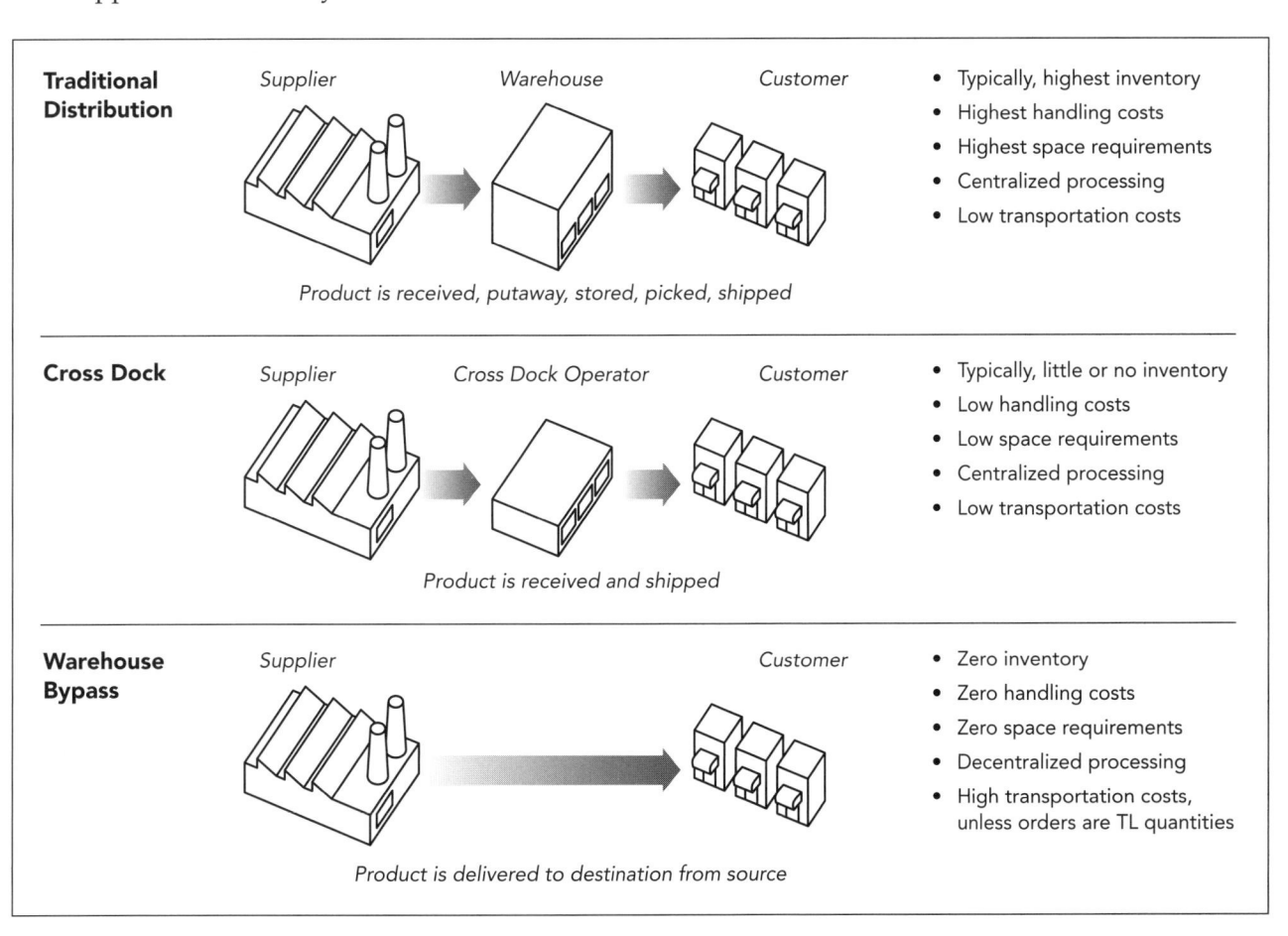

**Traditional Distribution**
Supplier → Warehouse → Customer
*Product is received, putaway, stored, picked, shipped*
- Typically, highest inventory
- Highest handling costs
- Highest space requirements
- Centralized processing
- Low transportation costs

**Cross Dock**
Supplier → Cross Dock Operator → Customer
*Product is received and shipped*
- Typically, little or no inventory
- Low handling costs
- Low space requirements
- Centralized processing
- Low transportation costs

**Warehouse Bypass**
Supplier → Customer
*Product is delivered to destination from source*
- Zero inventory
- Zero handling costs
- Zero space requirements
- Decentralized processing
- High transportation costs, unless orders are TL quantities

## FIGURE 7

**Sample Cross Dock Savings**

Labor cost savings/carton (from grocery pilot programs) . . $0.25

Savings if 8,000 cartons/day are moved . . . . . . . . . . x 8,000

**$2,000/day**

Labor cost savings (based on 250 operating days) . . x 250

**$500,000/year**
*(Money that goes back to the company)*

**If item previously had 6 weeks (0.115 year) of inventory:**

Avg. inventory in cartons based on demand of 8,000 cartons
. . . . . . . (8,000 x 250) x (6 ÷ 52)

231,000 cartons

Pallet positions (at 24 cartons/pallet) . . . . ÷ 24

9,625 pallets

Storage space savings (at $10/pallet position) . . x $10

**$96,000/year**
*(Reduction in storage space costs)*

Value of inventory (at $20/carton) . . 231,000 x $20

$4,620,000

Inventory risk cost savings (damage, obsolescence, pilferage, etc., at 3% of value/year, 0.115 year in inventory)
. . . . . . $4.62M x (0.03 x 0.115)

**$16,000/year**
*(Reduction in inventory losses)*

Opportunity cost savings of inventory (at 12% interest rate, 45 days paid inventory)
. . . . . (0.12/365) x 45 x $4.62M

**$68,000/year**
*(Savings in opportunity costs of inventory)*

**Possible annual savings!**

$500,000 + $96,000
+ $16,000 + $68,000

**$680,000**

---

2. **Cross Docking:** Product is received, processed (if needed) and shipped from a facility.

3. **Warehouse Bypass:** Product completely bypasses the warehouse or cross dock center and flows directly from the source to the destination.

One of the traditional roles of warehousing has been storing inventory in anticipation of demand. Although there is a strong trend toward reducing inventory-based warehousing, many companies still build warehouses to store products awaiting orders. In some cases, inventory is a necessity. Storage is often required to support manufacturing or post-manufacturing processes such as aging of alcoholic beverages and testing of electronic equipment.[8] Some level of inventory is also required for products with strong forward-buying opportunities, high seasonal variations, and unpredictability in demand. The challenge lies in finding and maintaining the most appropriate level of inventory for a SKU. Many companies have failed at this challenge and have exceeded their warehouse's capacities. They have turned to cross docking to reduce inventory.

At the other end of the spectrum is a distribution strategy involving the bypass of the field warehouse or DC—also known as *supplier direct, direct from manufacturer, drop ship,* or *direct store delivery* (DSD). In this strategy, product bypasses the warehouse and is delivered directly from a supplier's plant or warehouse to its end destination customer, plant, or store. Transit time is reduced because product does not have to stop at a customer's warehouse for storage or a cross dock center for additional handling, sorting or consolidation. In manufacturing, individual suppliers may ship component parts directly to the assembly line just at the time that they are needed for a production run. Component parts inventory is kept to a minimum. In grocery retail, this strategy is used for fresh-packaged bakery, dairy, and produce items, which are rotated on and off shelves within short intervals. The supplier not only delivers these products directly to the stores, bypassing the retailer's warehouse, but he may also control ordering, monitor inventory, physically load the shelves and remove all coded, unsold products.[9]

Despite advantages of bypassing the warehouse, firms still want to use cross docking. The reason: with various suppliers bypassing the warehouse and converging on a specific store or plant, the customer may experience the following conditions:

1. Increased costly activity at the unloading docks of the plant or store

2. Less control at the plant or store increasing the chance of theft

3. Congestion in the plant's yard and dock area or the store's parking lot and receiving doors

CHAPTER 1  *Cross Docking at a Glance*   19

4. Increased paperwork because of more individual supplier transactions

5. Increased inbound transportation costs

The best distribution is often a combination of strategies. The first step is to examine the present organization and the way business is conducted within and beyond the facility. The goal is to determine the appropriate level of throughput for each strategy: how much to cross dock, how much to store, and how much to deliver direct for a customer.

## 1.7   Benefits and drawbacks of cross docking

To explain the benefits of cross docking, consider the savings that some companies have already achieved. Figure 7 illustrates the projected savings a business may expect based on grocery pilot programs that show labor savings of approximately 25 cents per carton.[10]

With all of these benefits and bottom line savings, it is surprising most companies interviewed for this research admit to only cross docking full pallet quantities of 10 percent of their SKUs—a small drop in an ocean of possibilities. Figure 8 counters the advantages with some concerns and drawbacks.

Most of these drawbacks arise because many people are unfamiliar with the hows and whys of cross docking. This publication encourages logistics managers to take a closer look at this cost-saving strategy.

**FIGURE 8**

**Cross Docking Benefits Versus Drawbacks**

| Benefits | Drawbacks |
|---|---|
| • Speeds product flow and increases inventory turns | • Difficult to determine candidate products |
| • Reduces handling costs at the CDO | • Requires supply and demand synchronization that they cannot support |
| • Allows the efficient consolidation of products | • Less than perfect supplier relationships; little or no trust in suppliers; reluctance of suppliers |
| • Supports customers' Just-In-Time strategy | |
| • Promotes better asset utilization | • Union fears of losing jobs |
| • Reduces space requirements | • Inadequate facility and not enough of an investment return to justify purchasing a new one or changing an existing one |
| • Reduces product damage because of minimal handling at the facility | |
| • Reduces pilferage and shrinkage because of faster turnaround | • Inadequate information systems support |
| | • Management does not have a holistic and supply chain orientation |
| • Reduces product obsolescence and out-of-date conditions because product does not stay in the warehouse | • Requirement of fast turnaround causes concerns on the inability to check product quality |
| • Accelerates payments to supplier, thus fostering better supplier partnerships | • Fear of being out-of-stock with no back-up inventory |
| • Decreases paperwork associated with inventory processing | • Less expensive to buy in truckload quantities and warehouse them |

## 1.8 Applying cross docking

There is no one method that will work well for everyone. Cross docking has too many parameters that are often unique to a company or situation. The key is to systematically examine various aspects of an operation and determine how, where, and why cross docking should be implemented. The phases outlined below are guidelines through the process and are geared towards companies that are initially embarking on a cross docking program as a standard business practice. Four phases (Figure 9) are outlined.

### Phase 1: Assessing the potential for cross docking in an operation

Assessment begins with a detailed examination at how a company does business, the products carried, the suppliers, and the present system of information, transportation, operations, facilities, and equipment used to move products through the facility. This phase

**FIGURE 9**

**Making the Transition to Cross Docking**

**CURRENT DISTRIBUTION STRATEGY**

**PHASE 1  Assessment and Negotiation**
- Identify products and suppliers as candidates for cross docking.
- Identify weaknesses and strengths in current system: operations, facility and equipment, information systems, customer, and transportation.
- Develop preliminary recommendations to eliminate weaknesses and build on strength.
- Communicate recommendations with suppliers and negotiate preliminary cross docking guidelines.
- Determine cross docking throughput requirements for Phase 2: Planning and Design.

**PHASE 2  Planning and Design**
- Generate cross docking system designs.
- Perform an economic analysis on alternatives.
- Select the most appropriate cross docking system design.

**PHASE 3  Economic Justification**
- Create system-wide cost model and determine ROI.
- Create product cost model and determine product profitability.
- Determine projected cost and savings and level of sharing across the supply chain.

**PHASE 4  Implementation**
- Develop a comprehensive implementation plan.
- Implement a pilot program.
- Implement cross docking system-wide.
- Develop procedures and standards for periodically monitoring and expanding the program.

**CROSS DOCKING**

determines the network of products, suppliers, and systems required to initiate and sustain an effective and efficient cross docking effort. Products and subsequent suppliers are selected based on their cost effectiveness and other ease of implementing factors. Preliminary recommendations are presented to potential trading partners, and negotiations for sharing costs and benefits are initiated.

**Phase 2: Planning and designing a cross docking system**
There are various types of cross docking systems with varying levels of complexity. The most appropriate system design must be generated based on the products and suppliers selected and should incorporate the proper operations, information systems, facility, transportation, and equipment required to support cross docking. Where several alternatives are developed, an economic analysis of each alternative will be required, and the most feasible design can then be selected.

**Phase 3: Identifying and negotiating the costs and savings of a cross docking system**
Before any implementation can begin, the design must be economically justified so that management can provide the capital investment. There are two methods for identifying costs and savings. One method is developing a system-wide model to determine overall savings and costs, and the second method focuses analysis at the product level to determine the effect of cross docking on an individual SKU's profitability. With this collection of costs, the CDO can calculate true costs and savings that may or may not be shared with trading partners.

**Phase 4: Implementing and maintaining the cross docking system**
A positive economic justification of the proposed design paves the way for full implementation. It is important not to be overwhelmed by the details needed to start up a program, and a comprehensive implementation plan is required to ensure that key developments are not overlooked. Implementing a pilot program will enable a company to study the effects of cross docking on a smaller scale. The goal is to determine any weaknesses that can be addressed and eliminated before beginning full operations.

## 1.9 Trends that stress the importance of cross docking

Cross docking *by itself* does not solve the problems of the supply chain, but it is important to examine the current trends (new concepts, new technologies, changing philosophies) that rely on it to function more efficiently and achieve greater goals. JIT has already been discussed along with its impact on cross docking. Other trends include:

## Efficient Consumer Response (ECR) and Quick Response (QR)

ECR and QR are respectively the grocery and retail industries' strategy to shorten the pipeline from raw material to the point of demand at the checkout counter, thus accelerating replenishment and reducing inventory. QR emphasizes bar coding and scanning as fundamental. ECR does not. According to the 1996 Food Distributors International Productivity Analysis, the components of ECR are:[11]

- Electronic Data Interchange (EDI): a computer-to-computer exchange of information through standardized formats.

- Category Management: a method requiring the restructuring of buying or merchandising functions to operate and manage similar products as strategic business units.

- Continuous Replenishment Programs (CRP): demand-generated replenishment based on point-of-sale (POS) information.

- Flow Through: the use of cross docking techniques to decrease handling and reduce inventory in distribution facilities.

- Partnerships: the joint sharing of supply chain functions to maximize customer satisfaction. Included in this initiative is the promotion of vendor-managed inventory (VMI), where a supplier manages the inventory of its products and places orders on behalf of the retail customer.

These components work together in a *pull-based distribution system* where actual consumer demand at the point of sale automatically triggers needed replenishment of products. "Best practices" cross docking is based on a pull distribution system as part of a continuous replenishment program.

To demonstrate how cross docking fits operationally in these logistics trends, consider the following example, which is illustrated in Figure 10.

A cashier scans the Universal Product Code (UPC) of an item purchased by a mother of three at a grocery store. This single action triggers a flow of events that ripple throughout the supply chain. The point of sale (POS) scanning data is input into a perpetual inventory of the item in the store's computer. Determining whether replenishment or reorder of the item will be required is automatically made. Reorder points are based on rates of sales, category management guidelines, store demographics, and other relevant factors so that the proper inventory level is always maintained at the store. If the store's inventory level for that item is at or below the

CHAPTER 1   *Cross Docking at a Glance*   23

reorder point, computer-assisted ordering (CAO) techniques allow the store's computer to electronically transmit a purchase order via EDI for that and other items requiring replenishment. In a vendor-managed inventory (VMI), the order transmission goes directly to the supplier in real time. Meanwhile, the supplier uses the POS scanning data to streamline production planning processes and to reduce inventory. The supplier consolidates the products required by the retail store, pre-labels the order, and ships only the required quantity so that it arrives at the retailer's cross dock center on the day that a delivery for the grocery store is scheduled. At the CDO, the order for a specific store arrives, is cross docked and merged with other orders for that store. Because product moves based on actual demand, an inventory of the item is not maintained at the grocery chain's distribution facility. The whole replenishment process is part of the continuous replenishment process (CRP) undertaken by the chain. Cross docking is the most time- and cost-efficient method used to move the product through the facility and to support these initiatives.

## The Internet and e-commerce

The Internet is a vast network of mainframe computers linked by a standard communication protocol called TCP/IP (Transmission Control Protocol/Internet Protocol). This standard protocol allows computers with different operating systems to communicate with

**FIGURE 10**

**Illustration of Continuous Replenishment with Cross Docking**

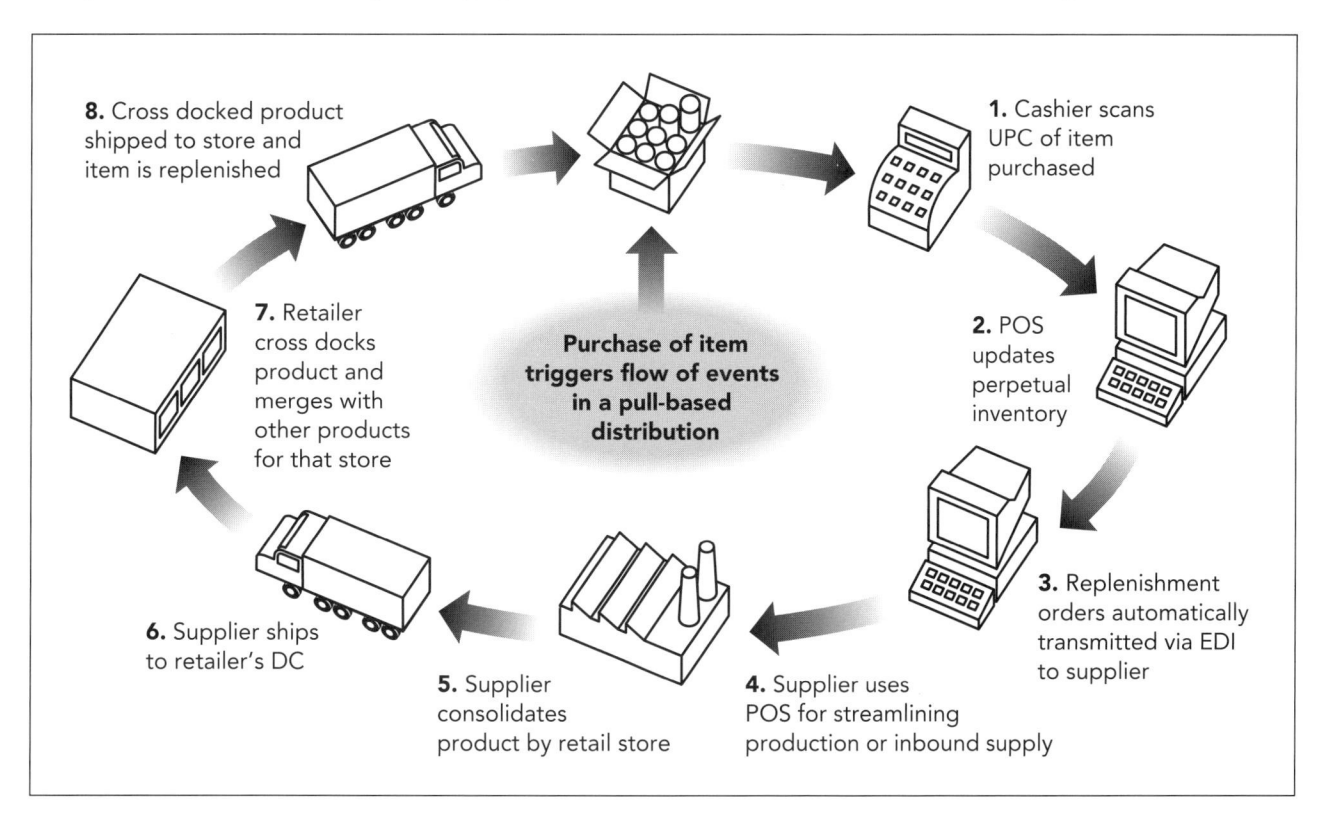

each other. Conducting computer-to-computer business transactions, or electronic commerce (e-commerce), over the Internet is a relatively recent, but rapidly expanding practice. Some manufacturers allow customers to view their order status, instantaneous inventory updates and orders at their Web site in real time. These manufacturers even allow customers to customize reports so that they can manage their inventories better. Smaller companies, who don't have the money to invest in EDI, can also transmit necessary information for cross docking over the Internet without the expense. Cross docking thrives on the timeliness and accuracy of real-time information so that product received can be immediately routed to its destination. It requires access to information from inventory, from orders, and from transportation beyond the four walls of the warehouse for product to move efficiently and not be stuck somewhere along the chain. Because of its real-time accessibility, the Internet will undoubtedly become a preferred method of communication for cross dock data in the future.

Another by-product of the e-commerce revolution has been the much-talked about growth in online shopping. In 1997, Web-based purchases amounted to $10 billion. By 2002, experts predict that Web-based transactions will reach $450 billion. In April 1999, Commerce Net and Nielsen media reported that 55 million people shopped online.[12] Online shopping is the fastest growing segment in the industry today, thus creating a tremendous impact on the cross dock operation of package delivery companies. At the end of 1999, UPS and Federal Express reported record increases in shipping volume, attributed to online orders. Intensive pre-season planning, hiring of additional workers, expanding operations and operating space, and adding extra trucks are some of the ways these companies were able to respond to recent surges in their business.

Online retailers, however, were not as ready. Toys R Us, Inc., for example, was unable to fulfill orders before Christmas, and offered $100 in store gift certificates to make up for their problems.[13] These problems include poor inventory management, slow order fulfillment systems, and not hiring enough people in their facilities. Because of the unpredictability and newness of this segment in the industry, formal cross docking programs specifically for online retailers are rare, if nonexistent. In the future, in response to customers' needs for fast service, order-fulfillment facilities for these Web-based companies may place specific high-volume products in the pick lanes in a continuous replenishment program supported by cross docking. Each time an item is picked for an online order, a perpetual inventory will be updated and suppliers immediately informed if replenishment is needed. Suppliers ship products to the fulfillment centers just at a time that they are needed to replenish the online retailer's picking modules.

**REAL WORLD EXAMPLE**

**Home Shopping Network, Inc., the Internet, and Supply Chain Execution Software**

Home Shopping Network Inc., based in St. Petersburg, FL, sells $1 billion in merchandise annually over television and on the Internet. It is currently implementing modules of a Supply Chain Execution software program by Optum, Inc. to allow end-to-end visibility from the time product leaves the manufacturer until it reaches fulfillment centers and gets shipped to customers. The company is planning to extend its information systems to the Internet so that customers can track the status of their orders and confirm delivery dates. (Source: Stein, Tom, "Optum lets users peek into logistics," *Informationweek*, May 3, 1999, p. 32.)

## Supply Chain Execution (SCE) Software

Over the past few years, Enterprise Resource Planning (ERP) companies have been buying Supply Chain Management (SCM) companies to integrate the business side of ERP with the logistics side of SCM.[14] In addition, WMS companies are buying Transportation Management System (TMS) companies—like the merger of McHugh Freeman and Weseley Software—to form McHugh Software.[15] It seems the software industry is merging or finding partners who will complement its software product and take better control of the entire supply chain. The goal is to create a Supply Chain Execution (SCE) software system that functions as an electronic medium for which multiple warehouses, plants, and DCs are integrated into an overall supply chain strategy. SCE deals with the seamless execution of supply chain activities required for the movement of the physical product through the pipeline—as opposed to SC Planning software that simply "plans," forecasts, or schedules these activities. SCE deals directly with what is happening on the floor.

With SCE, suppliers and CDOs can see what's being sold at the customer-end of the chain. With this information, what to ship can be more carefully planned, what to receive from a supplier can be more carefully planned, and the supplier can more carefully plan what to produce. This makes it possible to adopt transportation and inventory strategies that would have been difficult to implement in previous execution software, which consisted of systems that did not communicate with each other. Cross docking is one of these inventory strategies.

SCE's visibility with manufacturing, order management, transportation, warehouse management, and planning ensures that inventory for cross docking can be deployed based on the latest demand information. In addition, products can be made available not just from one facility, but from an entire network in response to unexpected changes in demand. It will eventually be more interactive so that the movement of products in transit can be redirected in real time to quickly satisfy customer demands, such as re-routing of delivery trucks and redirection of specific orders to certain facilities. Cross docking with SCE will enable the continuous movement of inventory, significantly cutting inventory carrying and handling costs.[16]

## 1.10 CD is here to stay

Cross docking is here to stay. The simple concept of shipping and not storing makes it an ideal solution to many logistics problems. To maximize its benefits, the program requires complete product visibility and information accessibility throughout the supply chain. Suppliers, CDOs, and customers must operate on a system based on cooperation, open communication and a cycle of continuous improvement. The details needed to make it work can be painstakingly numerous and sometimes complicated, but if successfully planned, designed, and implemented, it will lower operating costs, reduce inventory investment, improve product turns, and elevate supplier-customer relationships.

## Notes

1. Westburgh, Jesse, "Cross Docking in the Warehouse—An Operator's View," *Warehousing Forum*, August 1995, p. 1.
2. Schwind, Gene, "A Systems Approach to Docks and Cross Docking," *Material Handling Engineering*, February 1996, p. 59.
3. "Cutting Costs with Crossdocking," *WERCSheet*, September 1995, p. 1.
4. Cooke, James Aaron, "Cross-docking rediscovered," *Traffic Management*, November 1994.
5. Moore, Thomas and Roy, Chris, "Manage Inventory in a Real-Time Environment," *Transportation & Distribution*, July 1998, p. 70.
6. For more information on merge-in-transit, refer to Jones, Bill, et al., "Merge in Transit Works: We Can Prove It," *Annual Conference Proceedings*, Council of Logistics Management, San Diego, CA, Oct. 8-11, 1995, pp. 153-157.
7. White, John III, *Cross Docking Principles and Systems*, The Warehousing Short Course, presented September 9, 1996.
8. Shamlaty, Ron, "This is not your father's warehouse," *IIE Solutions*, January 2000, p. 32.
9. McEvoy, Kevin, "DSD or cross-docking—or both?" *Progressive Grocer*, March 1997, p.23.
10. Wagar, Kenneth, "Cross Dock and Flow-Through Logistics for the Food Industry," *Annual Conference Proceedings*, San Diego, CA: Council of Logistics Management, October 1995, p. 188.
11. Knill, Bernie, "Information Pulls Food Distribution," *Material Handling Engineering*, July 1997, p. 4.
12. Shamlaty, Ron, *loc. cit.*, p. 32.
13. Malloy, Amy, "Experience helps: E-shopping rises 270%," *Computerworld*, Jan. 10, 2000, p. 44.
14. Menezes, Joaquim, "ERP's future lies in supply chain," *Computing Canada*, April 16, 1999, p. 19.
15. Harrington, Lisa, "New tools to automate your supply chain," *Transportation & Distribution*, December 1997, p. 41.
16. From www.optum.com, official web site of Optum, Inc., which markets the SCE Response Center—an integrated supply chain execution solution for enterprises with multi-site distribution networks.

# Phase 1: Assessing the Potential for Cross Docking

**Objective of Chapter 2:** *To describe a procedure for assessing a company's capability to install and sustain a cross docking system.*

**2.1** Step 1: Review current corporate goals and trends

**2.2** Step 2a: Review products

**2.3** Step 2b: Review and select suppliers

**2.4** Step 2c: Audit current operations and policies

**2.5** Step 3: Analyze strengths and weaknesses

**2.6** Step 4: Develop short- and long-term recommendations

**2.7** Step 5: Quantify recommendations

**2.8** Step 6: Review recommendations with appropriate trading partners and negotiate the benefits and costs

**2.9** Step 7: Finalize recommendations

# Phase 1: Assessing the Potential for Cross Docking

For some managers, identifying the opportunities for a successful cross docking operation is not intuitively obvious. A company may be reluctant to pursue this technique because it cannot determine whether its business would be a likely candidate. Many questions arise: Are there any products that can be cross docked? How are these products selected? Are the vendors reliable? Is the current facility and technology appropriate for cross docking? These questions are critical to successful cross docking, but they are not difficult to answer. The key lies in analyzing current operations and systematically assessing whether a company has what it takes to cross dock.

Current literature is primarily about benefits and limitations, and provides many suggestions for implementation. However, few authors discuss how to assess the cross docking potential of a business. The common vein that many authors and practitioners share is the need to identify the "ideal" products to cross dock *and* to identify the suppliers willing to cooperate to ensure success. However, focusing only on products and suppliers can be shortsighted. To realize the full benefits of cross docking, assessment should begin at the top and stream down to the bottom. It should involve an overall examination from strategic, long-term goals to the operational, short-term objectives of facility and equipment planning.

The procedure outlined in this text will determine the CDO's capability to initiate *and sustain* a cross docking system. Variations of this approach may be required, with some steps added and/or eliminated, based on the nature of the business, resources available, and the specific goals that senior management wants to achieve. This approach is intended for the practical application of cross docking as a *standard business practice*—not as a quick fix or on an ad hoc basis. A conscious effort has to be made for a "phased-in" approach so as not to overwhelm the study team with too many suppliers or too many products, especially during an initial attempt to cross dock. In an interview, Virginia Carmon with IBM Global Services strongly suggests a pilot program starting with a few "win-win" suppliers and products and gradually adding more likely candidates as comfort level increases. Consider it a goal to constantly monitor changing business conditions and to determine the best combination of distribution operations that will provide the highest levels of customer service at the lowest overall costs. This combination of operations does not limit itself to cross docking. It should also include *warehouse bypass* and/or *traditional methods* of storing and shipping. The key is to identify the correct combinations and achieve a balance of service levels and costs.

For this research, an effort was made to provide a standard system that can be applied to *any* business (whether it is a manufacturer, retailer, wholesaler or third party provider). However, there are instances where specific issues relating to particular businesses are addressed. The entire process is illustrated in Figure 11.

## 2.1 Step 1: Review current corporate goals and trends

This step is structured to determine if there is a good foundation for cross docking. This foundation is primarily characterized by senior management's full support of the program and a willingness to provide the financial and labor resources to make the move to cross docking. It begins with a review of the corporate strategy by the entity that initiates the cross docking program.[1] It is at this step that project teams are formed and specific drivers propelling the need for improvements in current distribution operations are identified. Drivers may include reducing inventory assets, decreasing operating expenses, or decreasing cycle times. In this step,

**FIGURE 11**

**Phase 1: Assessment Process**

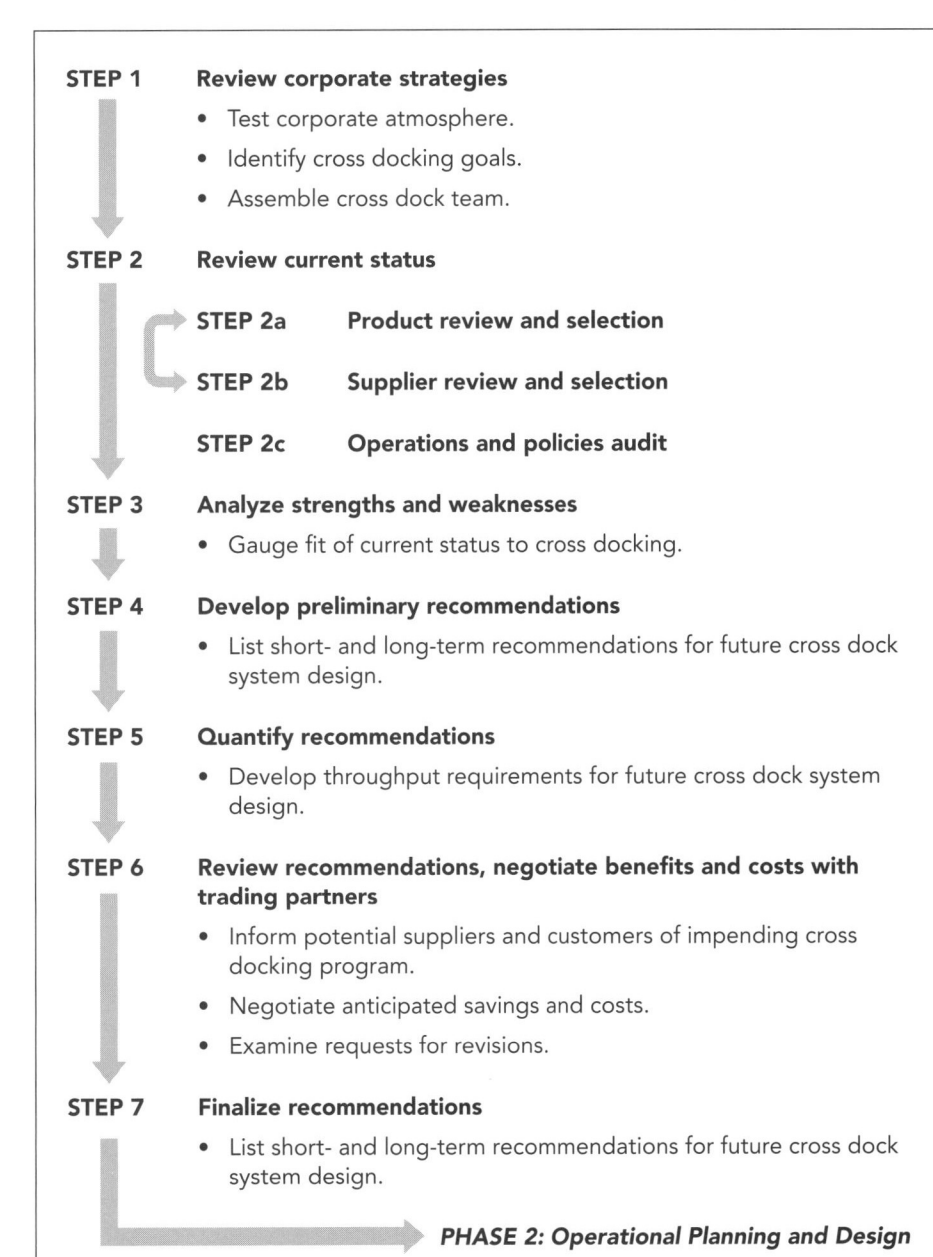

STEP 1  Review corporate strategies
- Test corporate atmosphere.
- Identify cross docking goals.
- Assemble cross dock team.

STEP 2  Review current status

    STEP 2a  Product review and selection

    STEP 2b  Supplier review and selection

    STEP 2c  Operations and policies audit

STEP 3  Analyze strengths and weaknesses
- Gauge fit of current status to cross docking.

STEP 4  Develop preliminary recommendations
- List short- and long-term recommendations for future cross dock system design.

STEP 5  Quantify recommendations
- Develop throughput requirements for future cross dock system design.

STEP 6  Review recommendations, negotiate benefits and costs with trading partners
- Inform potential suppliers and customers of impending cross docking program.
- Negotiate anticipated savings and costs.
- Examine requests for revisions.

STEP 7  Finalize recommendations
- List short- and long-term recommendations for future cross dock system design.

*PHASE 2: Operational Planning and Design*

senior management must determine if cross docking goals are consistent with corporate strategic goals and assemble the team that will undertake the analysis, planning, design, and implementation.

### Part 1: Consider the current corporate atmosphere

In this step, determination is made whether the company initiating the cross docking program is ready to support it as a standard business practice. The sidebar "Key Questions to Determine Cross Dock Readiness" lists questions to determine whether the corporate atmosphere is conducive to cross docking. An affirmative answer to each of the key questions indicates a good foundation for the program. A negative response does not necessarily spell doom, but it may limit the expected benefits.

### Part 2: Create a cross-functional supply chain team

Some experienced users interviewed for this research recommend forming a supply chain team first. This team would spearhead projects to continuously investigate supply chain initiatives, *not just cross docking*, in order to streamline the company's entire supply chain. Key players may include a Vice President of Logistics and managers of distribution, transportation, procurement, retail, and customer service with some involvement from managers of finance, information systems, sales and marketing. The supply chain team would oversee the cross docking project team.

### Part 3: Create a cross docking project team

This project team assesses the CDO's ability to sustain the program, develops designs, performs economic justifications, implements the strategy, and monitors the system. This team may include a project manager, the CDO's operations general manager, and a project engineer (for pricing in third party environments and for operations design) with periodic involvement from members of the supply chain team. After preliminary plans have been developed, representatives from suppliers and customers will join this team.

### Part 4: Identify cross docking goals and objectives

The project team identifies goals for cross docking and estimates planning horizons. Some examples of objectives include:

- Reducing inventory by 30 percent
- Reducing unit handling costs by at least 50 percent
- Shipping cross docked product within 24 hours of product receipt
- Eliminating direct store delivery and consolidating shipments
- Cross docking orders for slow moving items

## Key Questions to Determine Cross Dock Readiness

**Does your company recognize logistics as a major player in strategic goals?** In the "Global Supply Chain Management Benchmarking Study, 1998," supply chain managers, two-thirds from manufacturing and the rest from distribution, indicated that warehousing and transportation were seen as less important strategically to overall performance.[1] This is a disheartening revelation considering the best practice trend to recognize logistics as a major player in strategic goals. A worthwhile, perhaps over-used, example of the validity of this approach is Wal-Mart's phenomenal success. Wal-Mart builds customer loyalty by consistently providing superior customer service that is manifested in everyday low prices for readily-available merchandise. It is well-documented that their strategy of using logistics operations to create value for the business is what makes their superior customer service a reality. To them, logistics is a major component in realizing strategic goals.

**Does your company truly recognize customer's needs?** Today's customers want cuts in costs and cycle time, no stock-outs, excellent order accuracy, no damaged merchandise, and prompt deliveries. They also look for more value-added services such as floor-ready merchandise, custom-picked pallet orders, and labels on products. They expect these services at no additional cost; otherwise they will not remain your customers.[2] Cross docking addresses many customer requirements with particular emphasis on the following:

- Reducing costs due to a decrease in handling products
- Reducing cycle times because product moves faster
- Reducing damaged merchandise due to less handling
- Reducing inventory costs because storage is eliminated
- Reducing space costs because less inventory is stored

**Does your company have a strong urge to gain a competitive edge?** In this day and age, competition is fierce, and senior management must find ways to differentiate their company from the competition. Cross docking is such a differentiator. The market will reward those who can meet customer demand faster at a lower cost. By cross docking, product flows through the facility to customers in the shortest time possible and reduces operating costs, thus increasing profits.

**Does your company keep pace with technological advancements?** EDI, POS, and bar coding are some of the advances in information technology that have automated and improved logistics transactions. Implementing these advancements does more than set the stage for cross docking. It ensures accurate and timely information exchange and product visibility for all trading partners participating in the cross docking effort.

**Does your company want to improve its supply chain?** Companies are recognizing the value of looking at their businesses as part of a complex supply chain. A progressive company is constantly scrutinizing its supply chain to find areas where it can compress product delivery time and reduce costs. It expends a major effort to identify redundancies in the chain and improve the integration of supplier, distributor, and customer.

**Is your company pursuing improved business processes?** Manufacturers, wholesalers, and retailers who are at the leading edge of improved business processes are poised for cross docking. If you are in the grocery industry, are you pursuing ECR initiatives such as computer-assisted ordering, EDI, CRP, and category management? Have you implemented or are you seeking to implement ERP systems to help integrate the supply chain? If you are in retail, are any QR or VMI programs in place? Have you unlocked the potential of Activity Based Management? All of these initiatives work to provide the critical information and product visibility needed by each trading partner to cross dock in a timely and efficient manner.

**Is there a willingness to shift to a new way of thinking?** A company in search of world-class status cannot be timid in its pursuit of excellence. There must be a willingness to explore new paradigms and consider new techniques. To understand how common it is for companies to settle into a particular way of doing business, consider the grocery industry. Most grocers receive and ship meat and produce products in a 24-hour period. However, they still force the product through putaway, storage, replenishment, and picking, even though the products flow out of the DC on the same day. Cross docking, although simple in concept, requires experimenting with new methods of handling product and information.

A *yes* to these questions indicates that you have a good foundation for undertaking a cross docking program. Full sponsorship by senior management is the key. With it, one can start assembling resources to begin a cross dock study.

If you've answered *no* to some of the questions, you may not gain management's full support in a cross dock study. Therefore:

- Consider initiating changes to company philosophy and strategies.
- Communicate the importance of key goals to senior management.
- Re-assess the corporate atmosphere once strategic goals are in line with cross docking goals.

---

1 "Supply Chain Management isn't as good as managers think," *IIE Solutions*, March 1999, p. 10.
2 Andersen Consulting, *Warehouse Systems and the Supply Chain*. Illinois: Warehousing Education and Research Council, 1998, p. 4.

- Reducing cycle time from three days to one day for orders with cross docked items

Be wary of objectives that may conflict with each other because achieving one objective may violate another. In addition, objectives must be documented and communicated to all functions, and to senior management. A company planning to cross dock as a standard business practice will require more planning and operating resources than one looking for "quick-fixes." An objective of maintaining zero inventory and cross docking everything in the product line is rarely practical. Unless a company has unlimited resources, there will always be products, suppliers, distribution operations, etc., that are not ideal or economical for cross docking.

## 2.2 Step 2a: Review products

Step 2a consists of three important sub-steps that may be done concurrently and may require input from one step to another.

Products get cross docked. Selecting the appropriate products determines the type and magnitude of the system to be implemented. Some practitioners suggest that almost any item unloaded from a trailer can be cross docked. They contend that product characteristics should not determine the feasibility of cross docking, but rather that the business organization should create a network system that supports cross docking any item. Many users and experts disagree because there are items that are impossible to forecast that must be handled through traditional warehousing methods. For example, products with highly unpredictable seasonal variations require some inventory on hand. Some items do not have the handling characteristics that allow smooth and fast movement of material across docks, whether by lift trucks, pallet jacks, or conveyor systems. Cross docking implies fast efficient movement, and any hindrance to this movement will impede cross docking success.

There are other issues to consider for a comprehensive evaluation. The goal for this step is to identify the SKUs, prioritize them according to their suitability for cross docking and determine their appropriate handling methods. Figure 12 summarizes the steps for product review.

### Part 1: Collect the data for evaluating potential products to be cross docked

The following information will be required before a comprehensive evaluation can begin.

- For each SKU: *(Use the same unit of measure for all SKUs. This information should be accessible from a SKU master file.)*

CHAPTER 2   *Phase 1: Assessing the Potential for Cross Docking*

- Item number
- Description
- Price paid by CDO
- Cost per unit
- Special handling requirements (i.e., refrigeration, security, hazards, etc.)
- Dimensions (L, W, H) or unit cube
- Minimum receiving quantity (i.e., full pallet, floorloaded cartons, eaches)
- Minimum shipping quantity (i.e., full pallet, full carton, or eaches)
- Weight
- Units per carton
- Cartons per pallet
- Average inventory (for calculating inventory savings)
- Name of supplier
- Other special requirements such as irregular shapes, crushability and stackability or if special handling equipment is required, etc.

**FIGURE 12**

**Phase 1, Step 2a: Product Review**

**PART 1**  **Collect the data**

**PART 2**  **Analyze historical data and consider impact**
- Create the SKU order completion profile.
- Create the cube movement profile.
- Perform a shipment variation analysis.
- Merge and rank the items based on the profiles.

**PART 3**  **Analyze product characteristics and consider impact**
- Include qualifiers, other than historical data, that makes a SKU "cross dockable."

**PART 4**  **Analyze cost data and consider impact**
- Compare inbound and outbound freight costs, handling costs, inventory cost, and space cost if product was warehoused, cross docked, or bypassed the CDO.

**PART 5**  **Group according to logistics strategy and unit load characteristics**
- Which SKUs will be cross docked? Which will be warehoused? How many move in full pallet quantities? In carton quantities?

**PART 6**  **Summarize preliminary throughput recommendations**

- Customer/Store Order History. For each order:
  — Customer/store identification
  — Unique SKUs requested on the order and quantities of each
  — Order date
  — Specific facility processing order (if more than one)

A 12-month sample is necessary for highly seasonal businesses. However, if demand and movement do not vary significantly over the year, then a smaller sample might be sufficient. It is critical to focus on when peaks and valleys occur. If throughput does not vary significantly within a month, then monthly demand history for a 12-month period may be used, as a minimum. Be warned, however, monthly demand may not show beginning or end of month peaking.

- For cost benefit analyses, operational guidelines and costs are necessary including labor costs, cost of space, working days per year, working hours per day, cost of freight, inventory carrying costs, inbound and outbound freight costs, etc.

## Part 2: Analyze historical data and consider its impact

After the raw data has been collected, the next step is analyzing and questioning the raw data for input into the cross docking product model. Some of the statistics to determine for each SKU include:

- *Popularity.* This is sometimes referred to as "number of hits" or number of picks for each SKU. One with a high level of popularity implies that it is on many orders.
- *Demand (A).* This is the total quantity shipped during a specific period.
- *Unit cube (B).* This is physical size. It may be in a form of a carton, piece, or a pallet load, and its cube may be determined by multiplying length, width, and height. Another method for determining cube is to calculate the cube of an outer container divided by the number of pieces in the container.
- *Cube movement (A x B).*: This is the demand for SKUs multiplied by the cube for each one that equals the cube movement for the period. Handling requirements of an SKU that moves 50 units in a given time period may vary greatly, depending on the unit size—whether it is as small as a paper clip or as big as a refrigerator.

After this information is calculated, product profiles must be created. Profiling helps pinpoint products that possess the ideal conditions and characteristics for cross docking. It also determines trends for groups of SKUs with similar cube movement, popularity, etc. The following charts to profile the historical activity of SKUs should then be created.

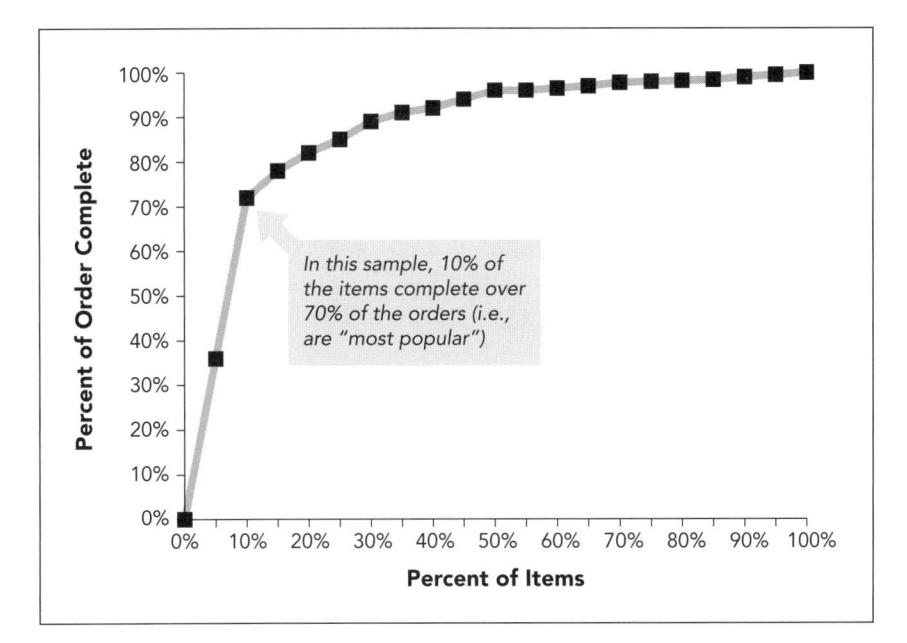

CHAPTER 2  Phase 1: Assessing the Potential for Cross Docking

**FIGURE 13**

**SKU-Order Completion Profile**

Adapted from Frazelle, Edward H., *World-Class Warehousing*. Georgia: Logistics Resources International, Inc., 1996, p. 60.

*In this sample, 10% of the items complete over 70% of the orders (i.e., are "most popular")*

- *SKU order completion profile.* A sample graph shown in Figure 13 indicates the percentage of SKUs that can complete a percentage of orders. This is determined by sorting SKUs by popularity. Starting with the most popular one, determine the percentage of orders that can be completed by picking that SKU. Progressively add more SKUs and determine what portion of the order is complete. In the example, a grouping of 10 percent of the SKUs can complete 70 percent of the orders. Cross docking the 10 percent items would mean reduced handling for 70 percent of the orders compared to traditional warehousing methods. To ease analysis, isolate and rank 20 percent of the most popular SKUs in this profile in a spreadsheet.

- *Cube movement profile.* (also classified as A Items) Most cross dock users cross dock the items with the highest cube movement. This analysis groups SKUs according to a specified cube movement range. With this profile, the percentage of SKUs that account for a specific percentage of cube movement is determined. Figure 14 shows a sample profile of cube movement. In this example, 16 percent of the SKUs account for 76 percent of cube movement. Again, isolate and rank 20 percent of the

**FIGURE 14**

**Cubic Movement Profile**

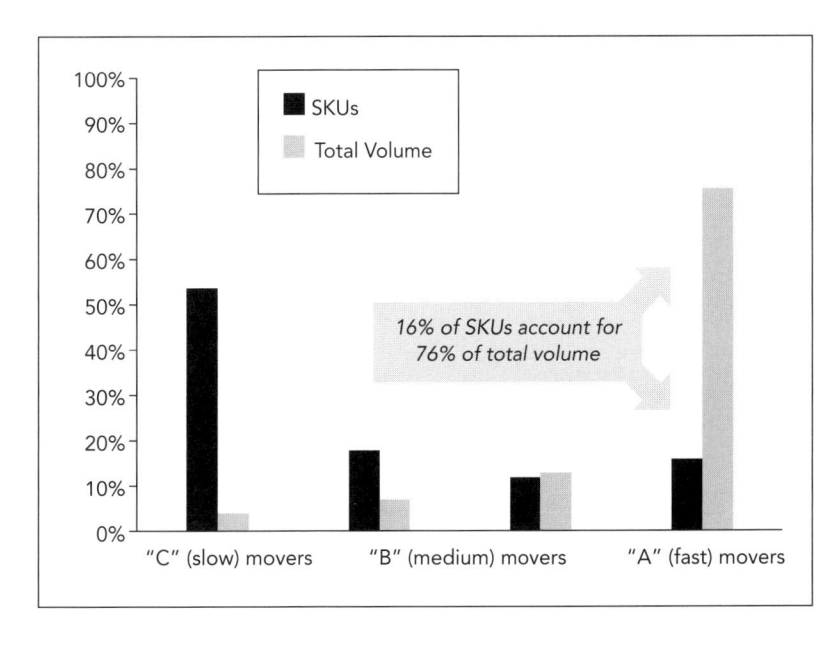

*16% of SKUs account for 76% of total volume*

highest cube movement SKUs determined from this profile in a spreadsheet.

- *Shipment variation analysis.* A product with erratic demand will be more difficult to cross dock than a more consistently demanded SKU. It becomes easier to synchronize inbound shipments to meet a demand that is fixed and known, rather than one that is unpredictable and erratic. SKUs with consistent aggregate shipping quantities have most of their values grouped around the mean and thus have a smaller standard deviation. Figure 15 illustrates this concept. To profile demand consistency, calculate the standard deviation and the mean of quantity shipped for each SKU. Calculate the ratio of the standard deviation to the mean, also known as the coefficient of variation. Suppliers of SKUs with the lowest coefficient of variation are likely candidates since their products are probably shipped in the most consistent demand pattern. Items with high standard deviations and coefficients of variation will probably need warehouse inventory to help handle volume peaks and valleys. Isolate and rank 20 percent of the most consistently demanded SKUs determined from this profile in another spreadsheet.

**FIGURE 15**

**Shipment Variation Analysis**

| PERIOD | Item A Vendor 1 | Item B Vendor 2 | Item C Vendor 3 |
|---|---|---|---|
| 1 | 614 | 2,791 | 1,432 |
| 2 | 1,073 | 3,086 | 1,903 |
| 3 | 145 | 981 | 1,415 |
| 4 | 717 | 3,208 | 1,874 |
| 5 | 696 | 2,651 | 1,458 |
| 6 | 614 | 2,418 | 1,150 |
| 7 | 832 | 3,445 | 1,777 |
| 8 | 691 | 2,792 | 1,853 |
| 9 | 513 | 2,127 | 1,281 |
| 10 | 899 | 2,974 | 2,268 |
| 11 | 578 | 2,501 | 1,480 |
| 12 | 304 | 1,872 | 1,111 |
| 13 | 787 | 2,479 | 1,511 |
| 14 | 657 | 3,062 | 1,839 |
| Mean | 651 | 2,599 | 1,597 |
| Std. Dev. | 224 | 608 | 318 |
| Coefficient of Variation | 34% | 23% | 20% |

Most Predictable Demand

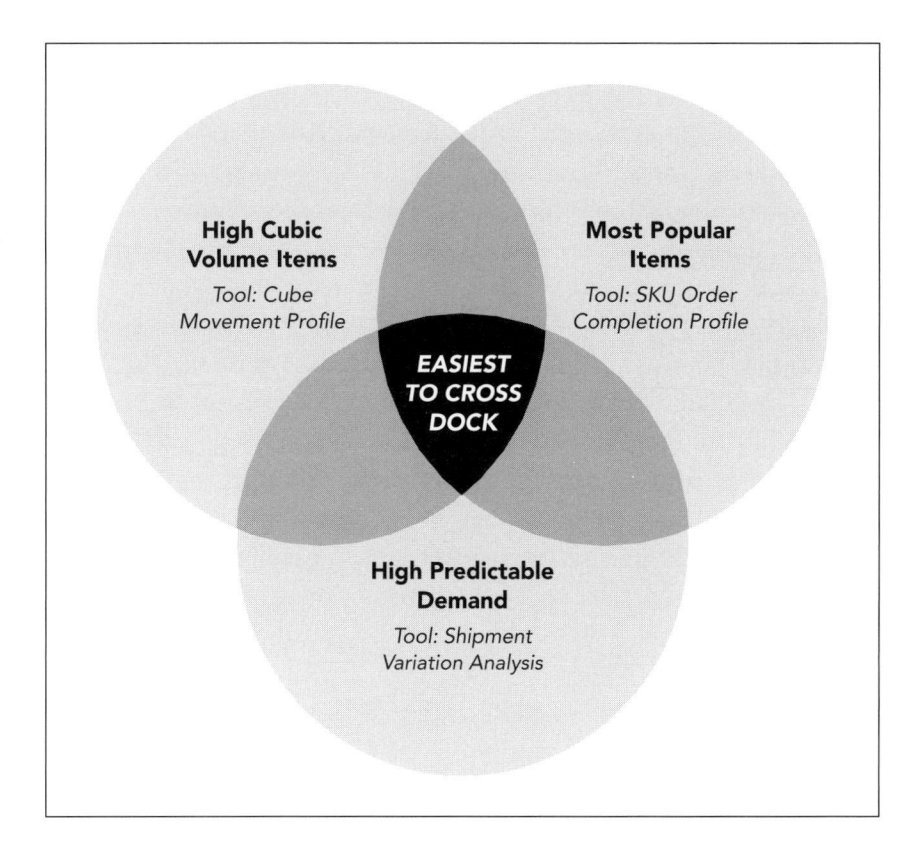

**FIGURE 16**

Product Selection Based on Historical Demand

[Venn diagram: High Cubic Volume Items (Tool: Cube Movement Profile), Most Popular Items (Tool: SKU Order Completion Profile), High Predictable Demand (Tool: Shipment Variation Analysis), with center labeled EASIEST TO CROSS DOCK]

- *Merge and rank the items from the above profiles.* Identify the SKUs that appear in all of the three spreadsheets and rank. Products with the lowest total ranking score would be the most ideal to cross dock. The Venn diagram in Figure 16 illustrates this concept. Products exhibiting high popularity, high cube movement, and high predictability are the best candidates and should be given immediate consideration.

### Part 3: Analyze product characteristics and consider the cross docking impact

Analysis of historical activity is a first-pass method for selecting possible cross docking candidates, but there are other qualifiers to consider to leverage the greatest benefits from cross docking. This analysis evaluates other product-related issues that may affect cross docking potential. Having one of these characteristics does not automatically categorize an SKU as "cross dockable." Instead it may be a combination of physical features and historical demand patterns that make it a good candidate for cross docking. The product characteristics that are critical to cross docking include the following items.

- *Unit load characteristics profile.* How are products received and shipped in the CDO's facility? Is it in single SKU pallet quantities, mixed SKU pallet quantities, floorloaded carton quantities, or less than full case (or broken case) quantities? Each type requires a different method of cross docking and implies varying costs and benefits to the supplier, CDO, and customer (Figure 17 illustrates these relationships) and, it appears that traditional single SKU palletized unit loads are the easiest and most economical to receive and ship. Lift trucks and pallet jacks are used to move these loads from receiving to shipping in a minimum amount of time and at a very low cost. However, avoid concentrating on only these types of products because doing so confines cross docking to a limited number of SKUs. Only a small number of customers' assembly plants and stores have the luxury of space to efficiently handle single SKU full pallets. At the other end of the spectrum, floorloaded carton and broken case quantities are more complex to cross dock without the proper sorting equipment. Cross docking these items may require an overhaul of the entire receiving and shipping process—unless a receive-sort-ship system is already in place.

**FIGURE 17**

**Unit Load Characteristics and Their Impact on the Cross Dock Operator**

| Unit Load Characteristics | Supplier | Third Party Provider | Cross Dock Operator | Customer's Retail Store |
|---|---|---|---|---|
| **Single SKU Pallets** | No incremental costs | N/A | Maximum productivity; Simplest to cross dock | May not have back-room or shelf storage available for large quantities of 1 item |
| **Multi-SKU Pallets and Display-Ready Pallets** | Incremental costs to assemble multiple SKUs on a pallet | May be used to assemble pallets from manufacturer | Maximum productivity; Simplest to cross dock | Direct move to selling floor |
| **Case Level (*Sort by stores or production run at CDO*)** | No incremental costs | N/A | Sortation required; high start-up costs; low to medium operational costs; complex to cross dock | If sorted according to store planogram, increase productivity; flexibility to order in small quantities; reduce expensive retail space costs |
| **Case Level (*Sort by stores or production run at supplier/third party DC*)** | Very high incremental costs if sort cartons in the facility. May be responsible for costs of letting third party provider do sortation | Sort cartons by store | Maximum productivity; low start-up costs; high operational costs since will be ultimately responsible for letting someone else do the sortation | If sorted according to store planogram, increase productivity; flexibility to order in small quantities; reduce expensive retail space costs |
| **Low Volume SKUs/Piece-pick** | Vendor keeps inventory of low volume/piece-pick SKUs. Only ship when ordered | May be used by supplier to hold low-volume/piece-pick SKUs | Eliminate low volume inventory from warehouse; cross dock low volume SKUs as soon as received from supplier or third party | Can order in less-than-carton quantities |

- *Handling requirements.* Handling requirements include issues pertaining to perishability, packaging, and lot control.
  - Highly-*perishable* items tend to benefit the most from a fast-flow environment such as cross docking. In addition, perishable goods often have high storage costs. Cross docking perishables may reduce these costs and ensure fresher products delivered to the stores.
  - Products in easy-to-handle *packaging* tend to be the most economical and the fastest to move—factors favorable for cross docking. Floor-ready items or products packaged for immediate display on the store floor is a growing trend. These items are easy to cross dock. But retailers are concerned that there is limited floor space for end-cap or end-of-aisle displays that can readily accommodate this type of merchandise. Although cross docking bulky items may decrease storage costs, it may also hinder fast movement of other products on the dock. Frequently these items require special handling and occupy valuable dock space, which will eliminate or reduce savings. An option would be to drop ship these large bulky items directly to customers' assembly plants or stores.
  - Ordinarily, products requiring *specific lot control* such as First In/First Out (FIFO) and Last In/First Out (LIFO) will complicate what may be an already complex system and should be dropped from consideration in an initial attempt at cross docking. Some companies, however, have redefined FIFO requirements in order to create a more flexible operation. Called *FIFO Granularity*, it involves defining a longer time window to specify the age of a product. For example, if FIFO granularity is defined for 6 months, a "new" incoming SKU will be considered equal to an "older" SKU manufactured 6 months ago or less and stored in the warehouse. Instead of retrieving the "older" SKU from storage, FIFO granularity allows the flexibility when cross docking items that just came in for immediate shipment.
  - Any inspection required on inbound items can compromise the speed of cross docking. Candidate items should have "built-in" quality instead of "inspected-in" quality.

**REAL WORLD EXAMPLE**

**Dell Avoids Inspecting Incoming Parts**

Dell Computer Corp., the Personal Computer manufacturer based in Austin, TX, has neither the time nor manpower—nor desire—to inspect incoming parts. Instead, it relies on regular on-site audits of suppliers, as well as quick, diagnostic tests during the assembly process. For critical parts, Dell can directly link itself to its supplier's manufacturing database. This allows Dell to monitor quality on its supplier's assembly line, and identify quality problems before products reach Dell. (Source: Minahan, Tim, "Dell Computer sees suppliers as key to JIT," *Purchasing*, September 4, 1997, p. 47.)

- *Need for value-added requirements*. Products that require ticketing, tagging, or hanging complicate cross docking. Time is needed for additional operations between receiving and shipping. With value added operations, cross docking crosses into the realm of flow through warehousing.[2] However, some suppliers offer display-ready merchandise as part of their product line. These items are received pre-tagged and pre-ticketed ready to be placed on the sales racks for immediate sale. These items are ideal to cross dock.

- *Importance to target customers*. The company may identify some items during certain periods that require fast flow through the facility because of their importance to target customers. Products "pushed" during promotional events and initial launches are conducive to cross docking because product is shipped without emphasis on demand planning. These may be items that may not have high demand, but are still ideal for cross docking.

- *Backorders*. Most companies recognize the urgency to cross dock backordered merchandise. These items are received, pre-allocated, and should be expedited immediately.

- *Part of replenishment programs.* Replenishment programs such as Efficient Consumer Response or ECR (for grocery) and Quick Response or QR (for retail) are considered "pull" distribution systems that traditionally use store and ship distribution techniques. But when used with cross docking, they provide the most significant impact to service levels and cost savings. Point-of-Sale (POS) data is scanned at the store and used in automatically triggering inventory orders for replenishment. Cross docking ensures fast and efficient expediting of replenished items in the supply chain.

- *Items with continuous, consistent demand*. Jan Young of Catalyst International points out that cross docking works best for SKUs that are consistently received and shipped daily. Component parts for continuous production manufacturing benefit substantially from cross docking. In the retail industry, "staples" such as milk, toilet paper, and diapers that have consistent, predictable demand are ideal. Historical demand can be analyzed using the steps previously described.

CHAPTER 2   Phase 1: Assessing the Potential for Cross Docking    41

- *Interstore transfers.* Forecasts are never perfect. Certain stores may be overstocked with too much of some items and understocked with others. For stores with overstocks, carriers on their return trip can cross dock the overstocked SKUs from one store to another store.

Evaluate SKUs based on these product-related issues: A rating system may be useful to narrow down the candidates. Some practitioners suggest concentrating on the top 10 to 20 percent of the most popular, highest velocity, and most predictable SKUs. This range depends on the time and resources available to the study team. This is a first-pass evaluation where the goal is to begin with products that will provide the easiest transition to cross docking. After further analysis, the cross docking team may decide to gradually evaluate more and more products as part of a strategic program to implement cross docking on a larger scale.

## Part 4: Analyze cost data and consider the cross docking impact

Products may exhibit historical data ideal for cross docking, but may not achieve any cost savings. This step involves using a cost model for an SKU (especially if one already exists) and quantitatively comparing the difference in costs for *traditional warehousing, cross docking,* and *warehouse bypass*. Some people suggest either a Direct Product Profitability (DPP) or Activity-Based Cost (ABC) model to determine the true cost of not putting away, storing, or picking an SKU. Developing these cost models is discussed in Chapter 4. For this first-pass analysis, a simpler cost model (Figure 18) illustrates a delivered cost comparison for a specific SKU. Similar comparisons are made for inbound and outbound freight costs, warehouse labor, inventory costs, and space cost of each SKU. For each SKU, select the most economical strategy.

Initially, it may take the most time in the entire assessment process to develop the evaluation methods in Parts 2, 3, and 4 of this product review. But once developed, these spreadsheets may be used repeatedly to allow analyses of the same products when business conditions change, or when other products are considered.

**FIGURE 18**

Delivered Cost Comparison for Candidate SKU

|  | Piece Cost | Inbound Freight | Outbound | Warehouse Labor | Inventory Cost | Space Cost | TOTAL |
|---|---|---|---|---|---|---|---|
| **As Warehouse** | $2.05 | $0.75 | $0.00 | $0.16 | $0.75 | $0.10 | $3.81 |
| **As Cross Dock** | $2.05 | $0.75 | $0.00 | $0.13 | $0.00 | $0.03 | $2.96 |
| **As Drop Ship** | $2.05 | $1.83 | $0.00 | $0.00 | $0.00 | $0.00 | $3.88 |

Most Economical

### Part 5: Group according to logistics strategy and unit load characteristics

After the above information has been evaluated for all SKUs, then group them according to logistics strategy (warehoused, cross docked, or warehouse bypass). For cross docked items, classify them according to their unit load characteristics. A cross docking system moving full cartons inbound and outbound is vastly different from a cross dock system moving full pallet loads. For each group of SKUs, sort and rank the potential for cross docking based on agreed-upon parameters. In the course of this research, some companies indicated that they begin with the SKUs with the highest cubic movement, highest popularity, and highest predictability. Then SKUs with product characteristics that are conducive to cross docking are added.

### Part 6: Summarize preliminary throughput requirements

After cross docked products have been identified, prepare a list of suppliers who will provide them. These suppliers will be reviewed in the next step.

## 2.3 Step 2b: Review and select suppliers

The next step is to identify suppliers who will provide the greatest probability for success in cross docking. A close working relationship between supplier and the CDO must be maintained. Any gaps in the relationship must be identified, addressed, and improved. Identification is achieved by evaluating prospective suppliers for the candidate products previously identified. Suppliers who have the capability to pre-allocate, sort and ship specific store orders to the CDO may seem ideal cross dock suppliers, yet a review of their performance will still be necessary to ensure that they deliver the correct quantity of the correct product on time and on schedule.

The following steps detail a supplier review for a fictional retail hardware company (Figure 19). Sample charts and worksheets illustrate the process.

### Part 1: If required, narrow down the candidates

If the CDO uses a few suppliers (fewer than 20), each one should be reviewed. However, in most

**FIGURE 19**

Phase 1, Step 2b: Supplier Review

| PART 1 | **Narrow down the candidates** |
|---|---|
| | • Focus on suppliers from top 100 candidate products from product review |
| PART 2 | **Gather supplier data** |
| PART 3 | **Calculate cross docking ease factor for each supplier** |
| PART 4 | **Evaluate other data specific to each supplier** |
| | • Is there a long and smooth trading relationship? |
| | • Does the supplier have the ability to enter into a collaborative relationship? |
| | • Does the supplier have the ability to shift to a new way of thinking? |

cases, the CDO deals with hundreds or thousands of suppliers, thus making the review of each one a tedious task. The intent is not to overwhelm the cross docking team. If there are more than 100 suppliers involved in the top 10 to 20 percent of the products qualified for cross docking (output from the product review), they may decide to review only the top 100 of those that are qualified.

## Part 2: Gather supplier data

Some issues must be evaluated for each vendor. The supplier data form (Figure 20, following page) may be used to gather information. If actual numbers are unavailable, a rating system may be used. In this evaluation, a rating of zero (0) represents the *best* performance and a rating of ten (10) represents the *worst*. Buyers, receiving supervisors, or personnel responsible for monitoring supplier shipments should administer this system.

The first questions relate to evaluating supplier reliability. This is one of the leading considerations of a cross docking effort. Erratic shipments by suppliers can lead to stock outs and customer dissatisfaction. On-time, accurate, and damage-free shipments must be the norm. The following information is tracked to evaluate reliability issues.

- *Record of a supplier's on-time shipments.* Frequency that a supplier delivers products on the agreed date and time for a specified period.[3] (Number of on-time receipts ÷ Total number of receipts = %)

- *Record of correct quantity received.* Frequency that a supplier delivers the correct quantity of an SKU. (Number of correct quantity receipts ÷ Total number of receipts = %)

- *Record of correct merchandise received.* Frequency that a supplier delivers the correct SKU. (Number of correct SKU receipts ÷ Total number of receipts = %)

- *Record of condition of merchandise.* Frequency that the overall condition of the merchandise is considered "acceptable" by receiving quality control personnel. (Number of acceptable receipts ÷ Total number of receipts = %)

Other quantitative information required to evaluate supplier candidacy for cross docking include:

- *Supplier shipping locations (state or country, if not the U.S.).* Suppliers that are physically close to the CDO have distinct advantages. Physical closeness reduces transit time and consequently lead time. In addition, close proximity fosters better integration of suppliers and the cross dock operator. It is more conducive for face-to-face contact, which is critical in cross

# FIGURE 20

**Form for Gathering Supplier Information**

## SUPPLIER INFORMATION
### FOR CROSS DOCK EVALUATION

Supplier Name: _____     Supplier Number: _____

*For each of the following categories, rate the vendor using a scale from 0 (Best) to 10 (Worst):*

|  | Best |  |  |  |  |  |  |  |  |  | Worst |
|---|---|---|---|---|---|---|---|---|---|---|---|
| On-time shipments | 0 | 1 | 2 | 3 | 4 | 5 | 6 | 7 | 8 | 9 | 10 |
| Correct/complete item quantity | 0 | 1 | 2 | 3 | 4 | 5 | 6 | 7 | 8 | 9 | 10 |
| Correct merchandise | 0 | 1 | 2 | 3 | 4 | 5 | 6 | 7 | 8 | 9 | 10 |
| Condition of merchandise | 0 | 1 | 2 | 3 | 4 | 5 | 6 | 7 | 8 | 9 | 10 |

*Please provide the following information:*

   Shipping Location (state, or country if not U.S.)   _____

   Lead Time (weeks)   _____

   Technology In Use (e.g., EDI? Bar coding? RF?)   _____

   Part of Label Compliance Program? (Y or N)   _____

*List item names and SKU numbers for this vendor: (indicate if promotional)*

_____     _____
_____     _____
_____     _____
_____     _____
_____     _____
_____     _____
_____     _____
_____     _____
_____     _____
_____     _____

*Continue on additional page if necessary.*

# CHAPTER 2  Phase 1: Assessing the Potential for Cross Docking

docking negotiations, provides a constant flow of information and communication, and helps in sharing resources such as technology for EDI. For this exercise, distances from suppliers were converted according to a scale agreed upon by the team and indicated in the sample supplier evaluation shown on Figure 21. In this example because the CDO is located in New Jersey, any supplier from New Jersey gets a score of 1 (almost excellent), while suppliers from the Far East get the lowest score of 10.

- *Supplier lead times.* This is the average length of time from placing a replenishment order to receiving the merchandise. To support a pull-based replenishment environment, cross docking works best with products that have short replenishment lead times. Because of the necessity for short lead times, suppliers physically closer to the CDO's facility will be better candidates than those who are far away. On the other hand, there is some merit for suppliers that have long lead times (such as imported merchandise). Plans for cross docking at the destination facility can be made while the product is still in transit. For this exercise, a pull-based environment is assumed and shorter lead times are desired. Average lead time is measured in weeks. For products with variable lead times from one supplier, specific lead times should be indicated next to the SKU.

**FIGURE 21**

**Supplier Evaluation**

**Supplier Name:** ........... LOCKS R US
**Supplier Number:** ......... 41699
**SKU Number:** ..... 12310934
**SKU Name:** ....... Cabinet 23 Drawer Plastic

*Rate the Supplier in every category from 0 (best) to 10 (worst)*

| CRITERIA | RATING | WEIGHT | GRADE |
|---|---|---|---|
| On-time Shipment | 2 | 1 | 2 |
| Correct and Complete Item Quantity | 2 | 1 | 2 |
| Correct Merchandise | 2 | 1 | 2 |
| Condition of Merchandise | 2 | 1 | 2 |
| Shipping Location (NJ=1; NY, PA, DE, MD=2; Midwest, Southeast=3; Southwest=4; West Coast=5; Canada=6; Mexico=7; Europe=8; Far East=10) | 5 | 3 | 15 |
| Lead Time in Weeks | 5 | 3 | 15 |
| Level of Technology | 2 | 2 | 4 |
| Weighted Sum | | | 42 |
| SKU Ranking from Product Review: (Refer to Sec. 2.2, Step 2a) | | | 41 |
| CROSS DOCK EASE FACTOR: (Product and Supplier) | | | 83 |

- *Current level of technology.* There is considerable debate about whether a high level of technology is expected from suppliers in order to ensure cross docking success. Some users believe that cross docking does not need technology, but relies on discipline among trading partners. Many still rely on phoning or faxing advanced shipping notices and orders. However, most users agree that technology improves the efficiency of operations, and they prefer to work with suppliers who keep pace with technology. These are the suppliers who have computer-to-computer information exchange capability, most notably the use of EDI with the transmission of ASNs. What is surely gaining ground is the use of the Internet to do the equivalent of EDI with regards to data transfer. Smaller companies that don't have the capital to invest in EDI can use the Internet for e-commerce with their customers.[4] Whichever method is employed, computer-based exchange of information increases speed, improves accuracy, and reduces costs. Other benefits include shorter and more dependable order cycles; decreased labor, freight, and material costs; improved cash flow; reduced inventory and improvements in overall business efficiency because complete, timely, and accurate information is readily available.[5]

For this exercise, a high level of technology was relatively important to the CDO. Suppliers that have EDI, provide ASNs, and use pallet license plates were given the highest scores.

**REAL WORLD EXAMPLE**

### American Freightways (AF) Keeps Pace with Cross Docking Technology

American Freightways, a less-than-truckload (LTL) company, based in Harrison, AZ, performs pure cross docking in 45 of 222 of its facilities, also called customer centers. The company uses the latest state-of-the-art computers, manufactured by Kinetic Computer Corp. of Billerica, MA, (www.kin.com) on dock doors to enable an efficient cross dock and provide the ability to track product. A worker scans a label and his user ID and the screen directs him to the appropriate outbound door. When a customer calls up inquiring about the location of a product, customer center workers can readily relay the information to him. According to Kinetic, the time workers spend on administrative duties has reduced by 30 to 40 percent. They predict that this technology will only get better, allowing real-time response in temperature-controlled conditions and complete connectivity between warehousing and transportation. The data electronically follows the product. (Source: Lear-Olimpi, Michael, "Looking for the fast lane," *Warehousing Management*, April 1999, p. 26.)

- *Label compliance program.* Suppliers that participate in label compliance programs are better cross docking candidates. In these programs, suppliers are forced by customers to label all products according to their specifications *before* they are shipped to distribution points. Cross docking cannot be carried out efficiently if products do not arrive labeled, tagged, or otherwise marked and identified. The CDO must never be required to open a carton to determine its contents. Labels should identify the contents of the carton or unit load. Using bar codes makes it possible to automate the data collection function. Bar coded labels have become "portable data files" that travel with the product. Some examples include the two-dimensional bar code symbology called PDF417 and MaxiCode. United Parcel Service uses the latter in high-speed automated cross docking systems without using a master database. Not only is the application of a label critical, but in fully automated systems, the label should be applied as early as possible for processing at high speeds.

- *SKU numbers and names of products from each supplier.* This is a list of items provided by each supplier.

## Part 3: Calculate the cross docking ease factor for each candidate supplier

Each category of information or rating should be weighted according to its probable impact on cross docking. The weights are agreed upon by the project team and those conducting the supplier survey in part 2. In our working example (Figure 21), supplier reliability was given the highest weight. The ratings are multiplied by the appropriate weights and added together. The weighted sum of each vendor added to the product ranking from the product review determines a relative *cross docking ease factor*. Vendors are then ranked from the lowest score (most likely to be successful in a cross docking program) to the highest (least likely to be successful). A sample data table is illustrated in Figure 22.

**FIGURE 22**

Summary of Supplier and Product Review

| Vendor Name | Vendor Number | SKU Number | SKU Ranking | (Wt. 1) On time Ship | (Wt. 1) Correct Qty | (Wt. 1) Correct Merch | (Wt. 1) Merch Cond | Loc State | (Wt. 3) Distance | (Wt. 3) Lead Time | (Wt. 2) Level of Tech | Wtd. Sum | Cross Dock Ease Factor |
|---|---|---|---|---|---|---|---|---|---|---|---|---|---|
| Akro-Mills | 12203 | 1523623 | 1 | 0 | 1 | 1 | 1 | OH | 4 | 2 | 0 | 21 | 22 |
| Ocram Ltd | 31404 | 1566623 | 6 | 2 | 2 | 2 | 2 | NJ | 1 | 2 | 2 | 21 | 27 |
| Mike Prod | 317602 | 1632252 | 3 | 0 | 2 | 2 | 2 | NC | 3 | 3 | 2 | 28 | 31 |
| Terry Ltd | 65912 | 1568225 | 2 | 1 | 1 | 2 | 2 | CA | 5 | 2 | 3 | 33 | 35 |
| Mike Prod | 317602 | 1235265 | 8 | 0 | 2 | 2 | 2 | NC | 3 | 3 | 2 | 28 | 36 |
| AL Ind | 551 | 1258232 | 12 | 3 | 2 | 2 | 2 | NY | 2 | 2 | 3 | 27 | 39 |
| AL Ind | 551 | 1652354 | 14 | 3 | 2 | 2 | 2 | NY | 2 | 2 | 3 | 27 | 41 |
| Barb Ltd | 14861 | 1225535 | 15 | 3 | 3 | 3 | 3 | OH | 4 | 3 | 3 | 39 | 54 |
| Mark Ent | 1604068 | 1364522 | 21 | 4 | 3 | 3 | 3 | NC | 3 | 3 | 4 | 39 | 60 |
| Akro-Mills | 12203 | 1246633 | 39 | 0 | 1 | 1 | 1 | OH | 4 | 2 | 0 | 21 | 60 |
| Locks R Us | 41699 | 1231093 | 41 | 2 | 2 | 2 | 2 | CA | 5 | 5 | 2 | 42 | 83 |

## Part 4: Evaluate other data specific to each supplier

This part of the supplier review investigates the quality of the existing relationship of the supplier with the CDO. It analyzes how well a particular supplier works with the CDO. Some suppliers may possess the qualifiers specified in the previous steps and yet not have the relationship or experience that will enable them to work well with the CDO. The CDO must ask the following questions and evaluate responses to perform a comparative analysis.

- Is there a *long and smooth trading relationship* with this supplier? The CDO should consider its history with a specific supplier. How many years have they worked together? A supplier who has had a long relationship with the CDO is usually a better candidate than a relatively new supplier. But the number of years together may not be as critical as their quality. Has it been a smooth relationship or were there many instances of differences and conflicts of opinion?

- Does the supplier have the ability to enter into a *collaborative relationship*? In some cases, cross docking may mean greater cost to the supplier because products received at the CDO's warehouse will have to be pre-assembled or placed in display-ready pallets. Thus, a considerable amount of negotiation may take place between the supplier and the CDO. To survive, there must be a collaborative relationship between trading partners—a willingness to work together towards a goal and in so doing share any increases in costs and savings. With cross docking, the goal is reducing costs and improving customer service for the *entire* supply chain.

- Does this supplier have the *ability to shift to a new way of thinking*? Mike Scott, Vice President of Distribution at Tops Markets, emphasizes one of the deciding factors used by his firm—*supplier sophistication*. By sophistication, he means that the supplier is not bound by the past, but can adapt to different methods of thinking. This supplier is usually, but not always, at the forefront of current logistics strategies and tools (JIT, Activity-Based Costing, etc.) and usually, but not always, able to adapt to higher levels of information technology (EDI, bar coding, RF, etc.).

This step in the supplier review is subjective and based on a CDO's actual experience and knowledge about a specific supplier. Potential candidates are prioritized as "most likely to succeed," "likely to succeed," "less likely to succeed," or "not likely to succeed." To start cross docking with "win-win" products and suppliers, start with "most likely to succeed" candidates.

## 2.4 Step 2c: Audit current operations and policies

After candidate products and suppliers have been identified, auditing current operations and policies begins. This step forces the CDO's project team to examine current methods of doing business and to pinpoint inadequacies in their systems. Some companies approach cross docking with an ardent, and often unsuitable, picture of implementing some specific "high-tech cross docking" system. It is fine to dream as long as the "dream" can be related to "reality." The only way to realize the dream is through examining, documenting, and understanding the current system—with particular emphasis on those issues that will be affected by cross docking (see Figure 23).

### Part 1: Operations documentaton

The operations review documents the current method of handling products and information in the CDO warehouse and identifies the functions that include:

- Receiving
- Inspection
- Putaway
- Replenishment
- Order Picking
- Staging
- Shipping
- Yard Management
- Value added activities such as repackaging, ticketing, and price marking

In process flow charts, details of current tasks or procedures must be documented for product movement and information exchange. Although putaway and replenishment may be eliminated in a cross docking program, it is still important to detail these steps so that all information transactions are addressed. In addition, cross docked items may have to be merged with warehoused items. Document inbound and outbound shipping schedules. Determine true costs of each activity, standard times, and the average frequency of performing each activity for each function for future cost benefit analyses and activity-based costing exercises.

### Part 2: Facility and equipment documentation

The facility and equipment review studies the physical facility where the CDO intends to cross dock. If cross docking will be done

**FIGURE 23**

Phase 1, Step 2c: Audit of Current Operations and Policies

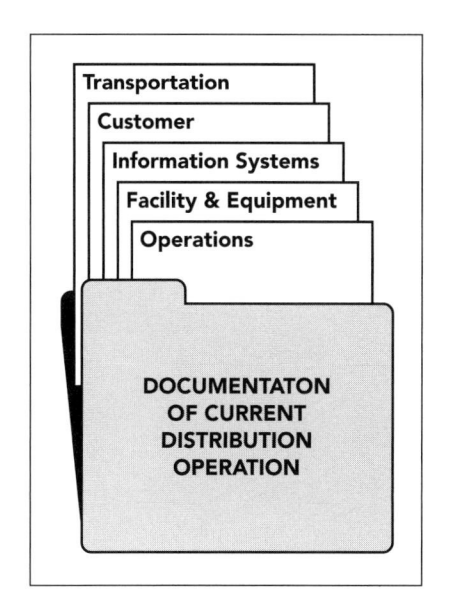

in a new facility, then the review would be limited to the equipment that will be re-used. Cross docking in an existing facility involves retrofitting the facility. For a third party provider, facility documentation will be based on the allocation of space for a particular customer. The following steps should be undertaken for a facility and equipment review.

- Obtain a layout of the current facility that includes the most updated locations of rack positions, permanent obstacles such as columns, fire protection systems, utilities, elevators, stairs, services, etc. It should also include locations of stationary equipment on the dock and accurate placement of dock doors.
- Track current products and equipment flows. Most facilities may have more than one type of product flow. For example, products that require refrigeration utilize a different flow from products that don't require temperature control.
- A cross dock program does not necessarily need sophisticated material handling equipment. Inventory available material handling equipment and their capacities, and document internal routings of material handling vehicles.
- Measure available dock space and staging space.
- Evaluate the docks and their condition. They literally set the stage for successful cross docking. Is proper safety equipment in place? Are the number of doors, lighting, heat, ventilation, etc. adequate? Is the dock floor in good shape?
- For future cost benefit analyses, determine the costs of space and equipment.

### Part 3: Information systems documentation

The information systems review examines the current functionality of the CDO's information system including:

- Audit the capabilities and limitations of the system documenting the WMS. Traditional warehouse management systems (early 1990s and older) were not designed to support cross docking and products are forced through inventory despite having the potential for cross docking. And a newer WMS may have been installed before cross docking was planned.
- Chart the current flow of information and methods of communication with the CDO's distribution center, the suppliers, and the customers. Consider the information flow for operational activities such as order management, order processing, distribution operations, inventory management, transportation, shipping, and procurement.

- Prepare an inventory of existing technology such as bar coding and radio frequency.
- Document the existence and functionality of other enablers such as ASNs and EDI.

### Part 4: Customer (store or assembly plant) documentation

This step documents policies and operations relating to receiving the cross docked merchandise. For retailers, these are the policies at the customers' stores or retail centers. For manufacturers, it would be the policies in receiving components at the assembly plant as part of a JIT program.

- Document schedules (dates and times) of product delivery.
- Measure service levels. This will be used to compare future service levels and service improvements.
- Measure available capacity for receiving merchandise. This may be the backroom storage areas of retail stores or the pre-assembly staging areas of manufacturing.
- Document ordering methods and lead time forecasting capability.
- Document the level of information systems technology.
  — Document methods of communication with the CDO and/or the supplier.
  — Document the existence and functionality of any cross docking enablers such as ASNs and EDI.

### Part 5: Transportation documentation

This step documents issues relating to the inbound and outbound transportation to and from the CDO's facility. If transportation personnel do not understand what warehouse personnel are doing, supply chain initiatives such as cross docking have little chance to succeed.[6]

- Measure inbound and outbound carrier performance for meeting scheduled deliveries.
- Measure inbound and outbound cube utilization.
- Determine inbound and outbound freight costs for cost benefit analyses.
- Document other methods of distribution such as direct shipping and how they integrate with traditional warehousing operations.
- Document methods and the protocol followed for carrier selection, scheduling, shipment consolidation and routing, shipment scheduling, and vehicle loading (e.g., Does the carrier use break bulk facilities? How many stops per trailer?).

- Document Transportation Management System (TMS) capabilities. Is it on a different computer system than the WMS? Investigate the ability to exchange information with the WMS.
- Document size of shipments and frequency of deliveries by customer segments.

## 2.5 Step 3: Analyze strengths and weaknesses

This next step is to identify where the current process *does* or *does not* comply with the vision for good cross docking practice. In this section, no attempt is made to list and discuss actual cross docking practices. The forum for that discussion has been reserved for the following chapter, which discusses operational planning and design. However, a short discussion of current practices and their impact on a cross docking program is included here (Figure 24).

**FIGURE 24**

Phase 1, Step 3: Analyze Strengths and Weaknesses

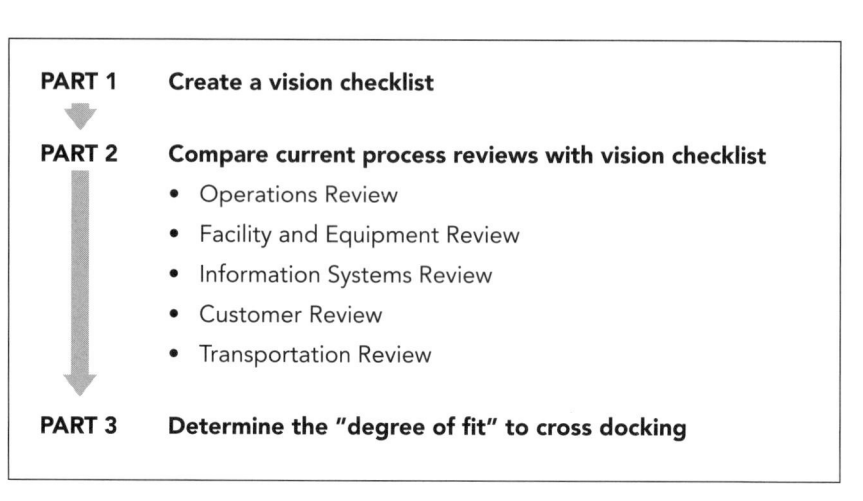

PART 1  Create a vision checklist

PART 2  Compare current process reviews with vision checklist
- Operations Review
- Facility and Equipment Review
- Information Systems Review
- Customer Review
- Transportation Review

PART 3  Determine the "degree of fit" to cross docking

### Part 1: Create a vision checklist

A vision checklist is a "wish list" of good practices, tools, and technology that will be beneficial for a future cross docking operation. The creation of a vision checklist identifies gaps between the current non-cross dock operation and the proposed one. Assembling this checklist requires research. Collect, read, observe, and document as many cross dock practices and operations as possible through site visits, case studies of actual operations, and articles and publications. Figure 25 illustrates a sample vision checklist.

### Part 2: Compare current process reviews with the vision checklist

At this stage, comparing "have nots" with "haves" is made easier by creating a checklist of conditions critical to cross docking success that may or may not be present in your business. A sample review table is provided for each function.

CHAPTER 2   Phase 1: Assessing the Potential for Cross Docking    53

# FIGURE 25

**Vision Checklist**

**Operations**
- Receiving/Staging
  —High dock capacity
  —Rapid rate of loading and unloading
  —High visibility of incoming receipts
  —Ability to schedule and confirm delivery
  —Ability to automate data collection (via bar coding, EDI)
- Inspection
  —Little or no inspection required
- Order Processing
  —Use of computer-based transactions and data transfers (such as EDI, Internet-enabled systems)
- Order Picking
  —Automated case level sortation
- Picking/Packing Verification
  —Automated processing
- Shipping/Staging
  —High dock capacity
  —Door-per-store
  —Ability to transmit shipments, schedule and confirm delivery to stores
  —Ability to automate data collection
- Yard Management
  —Tractor and trained driver available in yard
  —Yard manager available

**Facility and Equipment**
- Facility
  —Available dock space
  —Adequate number of doors
  —Good product flow
  —Facility not heavily racked
  —Good dock condition
  —Adequate trailer parking and yard space
- Equipment
  —Adequate conventional equipment
  —Automated sortation systems
  —Available pallet rack staging in dock area

**Information Systems**
- Warehouse Management Systems
  —Ability to send and receive ASNs
  —Ability to pre-allocate incoming receipts
  —Ability to confirm arrival time and date from carrier
  —Ability to receive order detail
  —Ability to select and schedule dock location
  —Ability to scan bar code on each pallet received
  —Ability to compare received pallet bar code to received ASN
  —Ability to identify and be notified of receiving variances
  —Ability to control sortation and other related equipment
  —Ability to create and track bar codes
  —Ability to track and report supplier and carrier performance
  —Ability to track and report warehouse performance
  —Ability to plan operations including manpower and dock utilization
- Radio Frequency Communication
  —Ability to process information in real time
- Bar Coding
  —Use of pallet license plates (UCC-128 or others)
  —Use of bar code readers and automatic scanning devices
- Electronic Data Interchange
  —Connectivity to the WMS
- Transportation Management System
  —Connectivity to the WMS
- Others
  —Interface to ERP systems
  —Productivity Tracking Software
  —Use of Activity-Based Management tools

**Customer (Store) Operations and Policies**
- Physical Facility
  —Available backroom storage
  —High capacity for display-ready pallets
  —Good dock conditions
- Operations and Information Systems
  —Automated ordering processing
  —Ability to identify CD availability for marketing events
  —Personnel available for restocking
  —Use of POS and computer-assisted ordering
  —Ability to monitor service levels
  —Ability to receive ASNs

**Transportation**
- Operations
  —Multiple stops per trailer
  —Drop trailers available
- Transportation Management System
  —TMS in place with WMS communication
  —Ability to track unloading/loading rates and other productivity tracking goals
  —Ability to communicate variances in order and actual shipments in real time

**FIGURE 26**

**Operations Review for Fictional Retailer**

- *Operations review.* Investigate the level of activity inside and outside the docks. With cross docking, more activity may be concentrated in these areas. Will your present operation be able to accommodate a marked increase in traffic, equipment, and personnel if required? Consider how information is processed. Automating operations and using computer-to-computer information exchange, especially when done in real time, can result

| Function | Vision Checklist | Yes | No | Observations | Impact on Cross Docking | Current Degree of Fit to CD — Low | Moderate | High |
|---|---|---|---|---|---|---|---|---|
| Receiving/ Staging | High dock capacity? | | X | Dock frequently congested | CD requires space on dock for rapid, unencumbered movement | X | | |
| | Rapid rate of unloading and loading? | | X | Excessive waits at receiving doors; took typically half a day to unload a full truckload | Any delays in inbound will create a ripple effect on the outbound | X | | |
| | High visibility of incoming receipts? | X | | Supplier transmits via fax to CDO incoming order; transportation routes and schedule deliveries | CD requires the advanced planning and scheduling of receipts to synchronize the timing of inbound deliveries and outbound departures. Can cross dock by faxing but ASNs are recommended for maximum productivity | | X | |
| | Ability to schedule and confirm delivery? | X | | Can provide carrier delivery time and carrier can confirm | CD depends on good communication, carrier reliability, and advance notification | | | X |
| | Ability to automate data collection (via bar coding, EDI)? | | X | Processing currently done manually, no scanning, no bar codes | May not be required for opportunistic CD, but bar coding preferred for large-scale cross docking | X | | |
| Inspection | Little or no inspection required? | | X | One in every 10 cartons currently inspected | CD assumes built-in quality—not inspected-in quality. Prefers that inspection be eliminated for approved suppliers | X | | |
| Putaway | | N/A | N/A | | Eliminated with CD | | | |
| Replenishment | | N/A | N/A | | Eliminated with CD | | | |
| Order Processing | Use of computer-based transactions and data transfers (such as EDI and Internet-enabled systems)? | X | | EDI capability between store and CDO; not all suppliers have EDI | Computer-to-computer communication among all trading partners not necessary, but preferred to minimize transaction costs and provide timely information for CD | | X | |
| Order Picking | Automated case level sortation in place? | | X | Manual, pick-to-pallet | Adequate for current volume but may be inefficient for large scale cross docking to an increased number of stores | | X | |
| Picking/Packing Verification | Automated processing? | | X | Manual check | CD may require a more automated system; Significant efficiencies readily achieved through bar coding and radio frequency | X | | |
| Shipping/ Staging | High dock capacity? | | X | Dock frequently congested | Though CD thrives for minimum "dwell" time, open space still required for increased shipping activity | X | | |
| | Door Per Store? | | X | Outbound pallets are staged on docks while awaiting cube-out volume | Dedicating a loading door for each store greatly facilitates CD as a trailer is always in place to receive material for that store. Eliminates the need for staging space and requires less handling of product | X | | |
| | Ability to transmit shipments, schedule and confirm delivery to stores? | X | | WMS provides ASNs to stores | Stores can be made aware of impending shipments; ASNs are recommended | | | X |
| | Ability to automate data collection? | | X | No bar coding | Fast-paced movement of products depend on automated methods to improve efficiency | X | | |
| Yard Management | Tractor and trained driver available in yard? | X | | | In CD trailers may have to be moved to meet a more rigid receiving/shipping schedule | | | X |
| | Yard manager available? | X | | Manages the yard | In CD someone needs to direct drivers so that trucks are spotted at the right doors. Personnel should be on hand to reallocate resources, work around problems, coordinate incoming and outgoing trucks and trailers | | | X |

in reducing paperwork and decreasing the time from receiving to shipping. Nevertheless it is *not* the main requirement in some cross docking operations. Cross docking of full pallets may be done with a clerk, a clipboard, and minimal coordination between receiving and shipping. Consider the unit load characteristics of the candidate products. It directly impacts the level of technology required for a successful cross docking program. Figure 26 is a sample operations review of a fictional retailer looking to cross dock and sort cartons to its stores.

- *Facility and equipment review.* Facilities that are small to medium in size may be heavily racked, or have limited open dock space, thus reducing the ability to cross dock efficiently. Many companies have yet to install the technological infrastructure to maximize cross docking in terms of mechanization and automation. The use of mechanized sortation systems at the case level greatly enhances cross docking operations. Although a sortation system may be expensive, it can provide the highest impact by eliminating inventory for a wider range of SKUs and improving service levels.[7] Figure 27 is a sample facility and equipment review.

- *Information systems review.* Cross docking is information oriented. Inventory is exchanged for information. Correct and accurate information is needed for what is being shipped in, when, and how. A CDO that has EDI capabilities and the ability to receive

**FIGURE 27**

**Facility and Equipment Review for Fictional Retailer**

| Function | Vision Checklist | Yes | No | Observations | Impact on Cross Docking | Low | Moderate | High |
|---|---|---|---|---|---|---|---|---|
| Facility | Available dock space? |  | X | No available space | CD requires space for rapid movement. Product review shows considerable case level movement thus requiring even more area for operations | X |  |  |
|  | Adequate number of doors? |  | X | Carriers kept waiting for open door | Expect more frequent inbound and outbound deliveries with proposed CD | X |  |  |
|  | Good product flow? |  | X | Warehouse is heavily racked; Long runs between receiving and shipping | CD thrives on smooth, unencumbered movement from receiving to shipping; the shorter the distance travelled between docks, the better for CD | X |  |  |
|  | Facility not heavily racked? |  | X | Warehouse is heavily racked | Exchange of storage space for dock space to succeed in CD | X |  |  |
|  | Good dock condition? | X |  | Adequate safety equipment, good floors, good ventilation | With CD, everything happens on the docks, thus the emphasis on good dock conditions |  |  | X |
|  | Adequate trailer parking and yard space? |  | X | Tight clearances for tractor trailer movement | No matter how well the CD system inside the facility, if trucks have difficulty accessing dock doors, then CD will be in jeopardy | X |  |  |
| Equipment | Adequate conventional equipment? | X |  | Adequate number of lift trucks, pallet jacks | Some CD products will be moved in full pallet loads and will require conventional equipment. If manual sort is still used, then pallet jacks will still be needed |  | X |  |
|  | Automated sortation system? |  | X | Currently sortation is manual | Product review show considerable carton movement in outbound shipping. Need further analysis. May require complex sortation equipment | X |  |  |
|  | Available pallet rack staging in dock area? | X |  | Used for staging | Allows for flexibility when performing CD |  |  | X |

# FIGURE 28

## Information Systems Review for Fictional Retailer

| Function | Vision Checklist | Yes | No | Observations | Impact on Cross Docking | Current Degree of Fit to CD — Low | Moderate | High |
|---|---|---|---|---|---|---|---|---|
| Warehouse Management System | Ability to send and receive ASNs from supplier? | X | | But many suppliers do not have EDI capability. Currently using fax | CDO should be notified of shipping time, date, carrier, SKUs, bar coding information of each order. Electronic exchange of information (such as via EDI or thru the Internet) is preferred for a more efficient CD but not necessary | | X | |
| | Ability to pre-allocate incoming receipts? | X | | Can note arrival of inbound goods and assign these goods to meet outbound orders | Especially useful in opportunistic CD. Material received can be directed to the staging area for consolidation into outbound trailer loads instead of storage | | | X |
| | Ability to confirm arrival time & date from carrier? | X | | Carrier can confirm | In CD, timing is critical. CDO must be informed of carrier arrival to schedule outbound shipment | | | X |
| | Ability to receive order details from customer? | X | | Stores transmit order details to CDO | CDO needs to plan outbound deliveries before inbound receipts for efficient CD | | | X |
| | Ability to select and schedule dock location? | X | | Can inform carriers of schedules and door assignments | Receiving and shipping carriers should be given the correct location at the correct time for efficient CD | | | X |
| | Ability to scan bar code on each pallet received? | | X | Currently no bar codes on pallets | The exchange of a large amount of information with one scan helps facilitate cross docking | X | | |
| | Ability to compare received pallet bar code to received ASN? | | X | Currently no bar codes on pallets | This allows the quick confirmation of supplier's shipment and the immediate dispatch of CD item to outbound docks | X | | |
| | Ability to identify and be notified of receiving variances | X | | Some manual entry in system currently required | CD requires a timely identification and notification of any receiving variances as it impacts immediately on outbound shipments | | X | |
| | Control of sortation and other equipment? | X | | Software available to communicate with the WMS | Especially important when case level sortation is required to stores | | | X |
| | Ability to create and track bar codes? | X | | But currently lacking equipment to print and apply bar codes; currently use paper to track | A WMS with the ability to create bar code and other label information for application to cases and pallets will help facilitate the downward flow of product from shipment to stores | | | X |
| | Ability to track and report supplier and carrier performance? | X | | Tracks on-time, correct, and damaged shipments | CD thrives with supplier and carrier reliability. The ability to track this will keep CDO informed of gaps in the supplier or carrier's performance | | | X |
| | Ability to track and report warehouse performance? | X | | Tracks productivity in each functional area | This assists in reporting improvements brought about by CD and identify system weaknesses; difficult to improve on something that hasn't been measured | | | X |
| | Ability to plan operations including manpower and dock utilization? | X | | Tracks manpower and dock utilization and suggests direction | With CD, everything happens on the docks, thus the emphasis on good dock planning and maximizing labor productivity | | | X |
| Radio Frequency Communication | Ability to process information in real time? | X | | Inventory is updated in real time with RF-based WMS | CD requires rapid, accurate exchange of information—especially when product has not been allocated upon receipt and will need to be allocated in real time at the receiving end | | | X |
| Bar Coding | Use of pallet license plates? | | X | Use of paper to track and direct product | (Same as above) | X | | |
| | Use of bar code readers & automatic scanning devices | | X | Use of paper to track and direct product | (Same as above) | X | | |
| Electronic Data Interchange | Connectivity to WMS? | X | | WMS has EDI capabilities | (Same as above) | | | X |
| Transportation Management System | Connectivity to WMS? | X | | Complements WMS system | Crucial that TMS and WMS communicate to coordinate truck and carrier operations with the operations within the CDO's facility | | | X |
| Other Information Systems Questions | Interface to Enterprise Requirements Planning Systems | | X | No ERP in place | Assist in supplier integration but not necessary for CD | | X | |
| | Productivity Tracking Software | X | | Provided by WMS | Helps in monitoring CD program and tracking savings and benefits. | | | X |
| | Use of Activity-Based Management tools | | X | Used for occasional analysis only | ABC can be used to build cost models and identify cross dock candidates | | X | |

and transmit ASNs with its suppliers may be in a better position for cross docking. In pilot tests conducted by the grocery industry, two distributors, one receiving health and beauty care products and the second, receiving paper products, used ASNs and UCC Code 238 pallet bar coding from manufacturers to speed receiving and to automate the process. ASNs alone enable CDOs to plan for receiving, sorting, and shipping, but when combined with pallet license plates, they eliminate visual inspection. The first distributor was able to document a $0.10 per case savings by using ASNs alone and this grew to savings of $0.25 per case when combined with the use of pallet license plates. The second found similar savings of $0.19 and $0.29 per case.[8] Figure 28 is a sample information systems review. In this case, the retailer had just installed a new WMS system but had yet to take advantage of bar coding in a manual operation.

- *Customer review.* Particular emphasis must be placed on available capacity of retail stores and production plants to receive merchandise. Determine if an adequate staging area is available for receipts. For retail, some stores have very limited positions for display ready merchandise. Figure 29 is a sample of a general store review by a retailer.

- *Transportation review.* Cross docking has been known to act as a "battering ram" against the wall that exists between transportation

**FIGURE 29**

**Customer (Store) Review for Fictional Retailer**

| Function | Vision Checklist | Yes | No | Observations | Impact on Cross Docking | Current Degree of Fit to CD |  |  |
|---|---|---|---|---|---|---|---|---|
|  |  |  |  |  |  | Low | Moderate | High |
| **Stores: Physical Facility** | Available backroom storage? |  | X | No available space | This makes CD more complex as inventory to stores must be kept at a minimum; entails more frequent deliveries at lower quantities | X |  |  |
|  | High capacity for display-ready pallets? |  | X | Only a handful of end-caps available | Display-ready pallets typically makes CD easier but if there is no available area for it at the stores then such a concept must be kept at a minimum | X |  |  |
| **Stores: Operations and Information Systems** | Automated order processing? | X |  | Use EDI to send order to distribution center | CD relies on the ability to exchange accurate and timely information |  |  | X |
|  | Ability to identify CD availability for merchandising events? |  | X | No process in place | Advance planning for promotional events facilitates the CD of promotional items | X |  |  |
|  | Personnel available for restocking? | X |  | Personnel dedicated to replenishing store shelves | Becomes critical to CD when backroom storage and docks need to be left open to absorb more incoming shipments |  | X |  |
|  | Use of POS and computer-assisted ordering? |  | X | Not currently | POS and CAO create a dynamic carton level CD environment. Order lead time is reduced, service level is increased and store's in-stock position is improved as demand driven orders drive the replenishment cycle. However, many operations can CD successfully without this luxury |  | X |  |
|  | Ability to monitor service levels? | X |  | Track on-time and complete order arrivals | Assists in monitoring effect (savings and benefits) of CD |  |  | X |
|  | Ability to receive ASNs? | X |  | Receive ASNs from CDO | Stores can be prepared for arrival of incoming shipments, but not critical for efficient cross docking |  | X |  |

**FIGURE 30**

**Transportation Review for Fictional Retailer**

and warehousing. Transportation should go beyond a closed world of tariffs and codes and examine where it can provide cross docking support. The Transportation Management System should communicate effectively with the Warehouse Management System to ensure efficient processing of cross docked orders. This is especially important when there are multiple stops per trailer. Without WMS and TMS connectivity, additional dock space will be required in the warehouse to sort store orders so that they can be loaded to suit the trailer's delivery route. But with WMS and TMS connectivity, TMS can relay the delivery route to the WMS, which in turn will release orders accordingly. Yard management and dock door utilization should also be reviewed. Cross docking impacts the flow and scheduling of trucks outside the facility. Priority should be given to cross docked items.[9] Figure 30 shows a sample transportation review.

| Function | Vision Checklist | Yes | No | Observations | Impact on CD (Cross Docking) | Current Degree of Fit to CD |
|---|---|---|---|---|---|---|
| | | | | | | Low / Moderate / High |
| Operations | Multiple stops per trailer? | X | | Trailer load can have multiple store destinations | This makes CD more complex as a store lay-down area will be required so that store orders can be held until they can be loaded to suit the delivery route. However, this issue can be eliminated if the TMS and WMS can sort the picking of orders to suit the delivery route | Moderate (X) |
| | Drop trailers available? | X | | Trucking company drops a trailer and loads it live all day | May increase costs for transportation but may provide flexibility on the dock and makes CD easier, less stringent, and may eliminate or reduce staging space | High (X) |
| Transportation Management System (TMS) | TMS in place with WMS communication? | X | | Connectivity with WMS | Transportation must work closely with warehousing to make CD work | High (X) |
| | Ability to track unloading/loading rates and other productivity tracking tools? | | X | Currently manually tracked by transportation supervisor | CD requires close synchronization of inbound deliveries and outbound shipments; delays in unloading and loading need to be immediately identified; automated tracking of productivity will aid in tracking carrier performance | Moderate (X) |
| | Ability to communicate variances in order and actual shipments in real time? | X | | Radio frequency in use to update order and create backorder status if order is not complete | Real-time exchange of information will prevent delays | High (X) |

## Part 3: Create a "degree of fit" table

Relate the fit of present conditions to cross docking success (Figures 26 to 30). The presence or absence of certain conditions dictates the high (H), moderate (M), or low (L) degree of fit of each component. This will assist the project team to focus its efforts for developing short- and long-term recommendations on those conditions with a rating of "L"—a low fit. To illustrate, a CDO with a heavily racked facility with no available dock space is given a "low (L) degree of fit" rating. The following step will address this issue and consider possible solutions.

## 2.6 Step 4: Develop short- and long-term recommendations

In Figures 26 to 30, weaknesses are indicated for those functions or categories that provide a low (L) degree of fit for cross docking. A list of criteria should be developed for prioritizing these weaknesses. This same criteria can also be used to determine which recommendations are short-term and which are long-term. Costs, available resources, technological capabilities, ease of implementation, importance to business or savings opportunity dictate the necessary planning horizons.[10]

**Sample short-term recommendations:**

- Recommend highly-qualified products and suppliers, identified in previous product-supplier review, for immediate cross docking.
- Approach suppliers for increasing product assortments and creating more display-ready merchandise.
- Ask suppliers to attach bar coded license plates to full pallet loads.
- Automate receiving and shipping processes using bar coded license plates.
- If allocation is known before the supplier ships the SKUs, approach suppliers about labeling cartons for certain SKUs with less than pallet order quantities.
- Improve dock utilization.
- Standardize store delivery schedules.

**Sample long-term recommendations:**

- Consider building a new facility or modifying existing facilities to maximize cross docking.
- Develop better order fulfillment processes and overhaul antiquated information systems. For standard cross docking system solutions, some experts estimate costs between $100,000 and $1 million. Companies seeking real-time inventory control and operating complex material handling systems should be prepared to spend anywhere from $250,000 to $2 million for the software package and integration with existing computer systems.[11]
- Develop better demand visibility by using the latest information system technology.

- Consider outsourcing specialty or low velocity SKUs to third party providers and eliminating them from the CDO's inventory. When orders for these low volume SKUs are generated, the third party provider assembles the orders and ships them to the CDO's facility where they are cross docked to their final destination.
- Adopt supply chain initiatives such as Quick Response and Efficient Consumer Response, if not yet in place.

## 2.7 Step 5: Quantify recommendations

This step finalizes throughput requirements to prepare for the next phase—the operational planning and design of the system. The following specifications are required:

- *Design year.* The design year is usually the maximum number of years that management can "accurately" forecast business trends. Some companies may have a four-year order backlog allowing them to forecast well into the future, and some companies can provide fairly reliable forecasts up to five years. Beyond five years, there may be a compelling need to totally reevaluate the operation. Based on the study's objective, the design year may also depend on the available capacity of an existing building if cross docking is to be retrofitted. However, if a company is contemplating a future property acquisition, the project team with senior management's guidance may need to determine how far into the future the new property can accommodate its business.[12]

- *Defined planning horizon.* Because cross docking can be a radical departure from traditional store-and-ship techniques, allowing a company to phase in cross docking over a defined planning horizon will help make the transition easier and smoother. Phase 1 may entail cross docking only 10 percent of the win-win, "easiest to cross dock" SKUs. This may involve a six-month planning horizon and a six-month implementation before Phase 2 is initiated. Phase 2 may increase the percentage of SKUs to 30 percent with a three-month planning horizon, and a three-month implementation. Phase 3, the final phase, may reach design year objectives of 60 percent of SKUs using a more automated system with a two-year planning horizon. With each phase, the project team carries over lessons learned from the previous phase, adjusts their design parameters, and proceeds to build a stronger system.

CHAPTER 2   *Phase 1: Assessing the Potential for Cross Docking*

- *List of cross dock products and their suppliers.*
- *Product characteristics.* These are unit load characteristics, maximum and minimum unit load dimensions and weights, and handling requirements for each SKU.
- *Projected throughput rate per planning horizon.* This is the forecasted cross docked product movement in and out of the CDO's facility for a specified time period. This should be grouped according to how much is moved in pallet loads or moved in cartons, pre-labeled or not labeled, conveyable or nonconveyable, etc.
- *Project inventory eliminated with cross docked items.* Cross docking SKUs will mean reducing or eliminating inventory for the same SKUs, and the storage space allotment will need to be recalculated for the facility re-design. Pallet rack positions may need to be exchanged for dock space.
- *Number of suppliers, supplier profiles, customers and customer profiles.* Supplier and customer profiles should indicate average and maximum cartons or pallets shipped per customer per day.
- *Recommended customer and supplier delivery schedules, patterns, protocol.* With a standardized delivery schedule, the cross dock team can create a throughput requirement of how many customers or stores to service in one shift. This will determine the number of units of handling equipment and labor required on the inbound and outbound sides. Suppliers and carriers must be involved in all team meetings involving scheduling decisions. Establishing delivery schedules is important for making sure that inbound items match outbound orders.
- If sortation is required, determine the following:[13]
  — Average and maximum volume throughputs for cartons requiring sortation
  — Sortation and merging requirements for mechanized systems
  — Method of communication with the WMS
  — Case label printing, application and verification requirements
  — Maximum and minimum carton dimensions of cartons requiring sortation
  — The location, placement tolerance, size, and symbology of existing bar code labels
  — Input and output buffer storage to increase labor productivity
  — Future sorting, capacity, and product requirements

## 2.8 Step 6: Review recommendations with appropriate trading partners and negotiate the benefits and costs

The key to effective relationships with trading partners is communication. In this step the CDO communicates its initial cross dock plans to potential suppliers and customers early in the planning process. This is also the beginning of what can be a long and intense period of negotiations. By involving trading partners early, a working relationship can be established and impediments to implementation can be identified immediately and addressed. It makes no sense to pursue a cross docking program of product assortments if suppliers will refuse to provide them.

The key question that a CDO must ask its suppliers and customers is: How can we work together to reduce costs and increase savings?

A key issue that is gaining ground in the supply chain is "gain-sharing" or working together as a team to help each trading partner in the chain realize savings to achieve mutual gain. The same focus should apply to cross docking. Cross docking players must recognize how the program translates into net landed cost savings. Nevertheless, the sharing of these costs and benefits can be extremely complicated—especially between supplier and CDO. *The difficulty of these negotiations must not be underestimated*. Companies interviewed for this research talked about the magnitude of negotiations that had to be accomplished—especially when the goal is for the supplier to custom-mix, pre-label, and pre-sort the pallets. It is important that cross docking does not transfer these costs to the supplier. Cross docking should be employed only in supplier-CDO-customer relationships where the overall supply chain benefits outweigh added supplier costs.

The following guidelines can aid in negotiating. These are a sampling of methods used in industry today. Examples have been provided to illustrate the concept.

The CDO must convince suppliers that cross docking will be able to:

- *Reduce overall supply chain costs.* Supplier and CDO should get together and see each other's costs and determine how cross docking can reduce them. These costs should be designed for both sides to "open their books" and understand the costing process, especially the need to make a profit. The key is to keep the reporting of these costs simple enough for both parties to compare them. No one likes a complicated spreadsheet with meaningless numbers.

- *Increase product marketability.* Suppliers have to be convinced that getting more product displayed more frequently drives enough incremental value and profit to offset the added cost.[14] Procter & Gamble (P&G), one of the leading manufacturers in the industry, encourages its retail chain customers to cross dock. They recognize that the savings that their customers gain from cross docking can be "reinvested" into improving the marketability of their products, thus ultimately increasing P&G sales revenue. The complication lies in that it is often difficult to quantify this increase in revenue.[15]

- *Increase trust in a supplier.* Developing trust is difficult for many supplier-CDO relationships. In some cases, there is an almost adversarial relationship. According to Virginia Carmon of IBM Global Services working together enables suppliers to attain a more favorable status with their customers. Customers will tend to do *additional* business with suppliers that support cross docking than those who don't.

**Hills Department Stores Works With Suppliers**

REAL WORLD EXAMPLE

Hills Department Stores and the Hartz Mountain Corporation, one of their suppliers, have a unique partnership built on trust and cooperation. Previously, a merchandiser from Hartz would visit Hills stores on a weekly or bi-weekly cycle for the purpose of writing the order, delivering the order, checking in the order, and cleaning, straightening, and adjusting the planogram. Presently, Hartz uses daily POS data provided by Hills to replenish the stores. Now the supplier maintains a perpetual inventory by item, by store, with reorder points and quantity based on seasonality and promotional activity. Actual sales update this inventory and create orders automatically. These orders are transmitted to Hartz who then ships the pre-allocated quantity to Hills cross docking system using bar coding (UCC/EAN 128) in conjunction with an ASN through EDI. Hills pays Hartz based on products which are scanned in pallet or case quantities and cross docked to their stores. By putting their customer's POS data to use, Hartz has been able to offer Hill better prices on products while reducing total inventory, reducing inventory turns on products, increasing profit and improving return on investment. (Source: Rouland, Renee Covino, "Perpetual Partners," *Discount Merchandiser*, December 1994, p. 24.)

- *Provide cash-back dollar amounts to suppliers.* "Money talks." Carmon also suggests that if a supplier, despite incurring additional costs, decides to participate in a cross docking program, then the CDO might decide to return some savings to the supplier by providing a cash-back dollar amount if suppliers label and consolidate orders for stores. The actual cash-back dollar amount is best determined after Phase 3: Identifying the costs and savings of a cross docking system (Chapter 4).

- *Improve overall savings or suffer the consequences.* In some cases, cross docking is simply a matter of power. Mass merchandise retailers state, "If you want to do business with us, it has to be done our way." Chargebacks are used by some retailers against suppliers who the retailers feel should be able to comply with their requirements. For an anonymous department store retail chain, the chargeback to a supplier is $25 plus ten cents for each unit shipped without a UPC. On a more rigorous level, the chargeback penalty is $5 per carton without an ASN and an additional $5 per carton without a bar code or UCC 128. The mass merchandiser Target requires that all soft goods be floor ready upon receipt with items placed on the approved hanger, folded to their specifications, with the right ticket.[16]

- *Improve their level of technology.* Some retailers share the technology and knowledge gained from cross docking with more sophisticated suppliers to their smaller suppliers. Wal-Mart often works with smaller suppliers to help them come up to speed.[17]

- *Share costs.* When a supplier sees his costs go up by twenty-five cents a carton, but the CDO's costs go down fifty cents a carton, the supplier should consider asking for another twenty-five cents per carton. The supplier breaks even, while the CDO still saves twenty-five cents per carton. Costs are shared, but savings are still achieved.[18] Determining the actual costs to be shared is best achieved at the end of Phase 3: (Chapter 4).

- *Pay suppliers faster.* Suppliers should be informed that with cross docking and decreasing the order cycle time (from order placement to delivery of merchandise), payment for merchandise can be quicker than for traditional warehousing. Hills Department Stores and the Hartz Mountain Corporation, one of their suppliers, work together so that the retail chain is automatically charged for each carton scanned in their cross dock operation.

The CDO must convince customers that cross docking will allow it to:

- *Pass savings on by offering lower prices to customers.* Customers must be convinced that cross docking can lower overall costs and be willing to pass those savings to consumers through lower-selling prices.

- *Reduce on-hand supply in stores and assembly plants.* Space is expensive. Customers should understand that with cross docking, they can order in lower quantities to reduce backroom storage and staging space.

- *Entice target customers.* Retail store managers must order cross docked product and believe that it can reduce store labor hours and entice customers through favorable prices.
- *Use lower prices on cross docked merchandise.* Customers' merchandising and marketing divisions can use lower-priced cross docked items to complement marketing objectives.
- *Maintain their way of doing business.* At the store level, Tops Markets made the shift from direct-store-delivery (DSD) to cross docking with some changes at the store level. Stores had to adapt to having the product come in at a different time of the day. Some stores complained because, with DSD, the suppliers restocked the shelves. With cross docking, store personnel had to restock the shelves. This was resolved by having the supplier send their employees to the store by car at the time of receipt. The CDO and the customer worked together to form a mutually agreeable arrangement.

Mass merchandisers have paved the way so that more and more suppliers are familiar with the support system required for cross docking. Most suppliers already understand the savings to be made and actively participate in the program with little or no negotiation. The difficulty lies in suppliers who are fairly new to the concept. These are the suppliers that the project team could spend time negotiating for cross dock support. If it is difficult to negotiate with a supplier, then he might as well be eliminated from the program. Cooperative alliances makes cross docking function easier and smoother.

Consider the following guidelines when communicating to suppliers about the components of a cross docking project:[19]

- A long-term true partnership is the goal.
- Benefits, requirements, and measurements must be clearly articulated.
- Requirements documentation must be stable, not a moving target.
- There must be a defined contact list of a *minimal* number of people.
- There should be ongoing meetings and reviews of progress.
- A cross-functional team from logistics, systems, and procurement should be assembled.
- At startup, (1) communicate well-defined requirements, (2) set and agree on a review process, and (3) determine and agree on key performance indicators.

After preliminary negotiations with the customer, revisions in throughput requirements, such as adding or removing product candidates and eliminating suppliers from further consideration, may be required.

## 2.9   Step 7: Finalize recommendations

*Any revisions to the recommendations and quantitative requirements should be agreed upon and documented by the project team and the concerned party.* Then the study team must present a final list of recommendations with expected planning horizons for the next phase of making the move to cross docking–*Planning and Design*.

## Notes

1. The initiator of the cross dock strategy may not always be the CDO. For example, third party providers who execute the strategy seldom initiate it. It is their customers who may call for cross docking to be implemented.
2. See Chapter 1.
3. The definition of "late" shipments will depend upon each operation. Operations with rigid truck dock schedules may consider that a truck is "late" if it is not at the designated dock within 30 minutes after its scheduled arrival. Others might provide one to two hours of leeway.
4. See Chapter 1 for a more detailed discussion on the Internet's impact on logistics and distribution.
5. Daugherty, Patricia J. "Strategic Information Linkage," *The Logistics Handbook*, New York: The Free Press, 1994, p. 765.
6. Andel, Tom, "Efficient Transportation Starts in the Warehouse," *Integrated Warehousing & Distribution*. June 1998, p. 84.
7. Wagar, Kenneth. "Cross Dock and Flow Through Logistics for the Food Industry," *Annual Conference Proceedings*. San Diego, CA: Council of Logistics Management, Oct. 1995, p. 186.
8. *Ibid*. p. 188.
9. Andel, Tom. "Efficient Transportation Starts in the Warehouse," *Transportation & Distribution*, June 1998, p. 88.
10. "Cross Docking: Process for Success," *Special Report for FMI/GMA Replenishment Excellence Conference*, New Orleans, LA: Sept. 1995, p. 5.
11. Cooke, James Aaron. "Cross-docking software: Ready or not?" *Logistics Management*, October 1997, p. 58.
12. Napolitano, Maida, *Using Modeling To Solve Warehousing Problems*, Warehousing Education and Research Council, Chicago, 1998, p. 54.
13. Schaffer, Burt. "Implementing a Successful Cross Dock Operation," *IIE Solutions*, October 1997, p. 36.
14. Jim Blaser, *loc. cit.*, p. 9.
15. Novack, *loc. cit.*, p. 152.
16. Norek, Christopher and Wright, Jimmy, "Working With Mass Merchants to Achieve Mutually Beneficial Supplier Requirements," *Council of Logistics Management Annual Conference Proceedings*, Florida, Oct. 20–23, 1996, p. 497.
17. Norek, *loc. cit.*, p. 499.
18. Cockerham, Paul W. "Throughput Rules Distribution," *Frozen Food Age*, July 1998, p. 21.
19. White, John III, *Cross Docking Principles and Systems,* The Warehousing Short Course, September 1996.

# Chapter 3

# Phase 2: Planning and Designing a Cross Docking System

**Objective of Chapter 3:** *To develop alternative physical, operational, and information systems requirements for a cross docking system, using the recommendations and throughput requirements developed in Chapter 2.*

**3.1** Types of cross docking systems

**3.2** Step 1: Generate cross docking system design

**3.3** Step 2: Perform an economic analysis on the alternatives

# Phase 2: Planning and Designing a Cross Docking System

With the determination of *what* to cross dock, the next logical step is to determine *how* to cross dock. However, before designing a specific system, the study team needs to get acquainted with the types of cross docking operations in a distribution environment. By knowing these operations, they will be better equipped to find the best system solution for their unique operation. Undoubtedly, there will be more than one solution. Multiple combinations of equipment, operations, layout, and information systems produce a wide range of throughput expectations, savings, and costs. For each viable alternative, the team has to calculate space, handling, and other related requirements. After these are identified, the team can proceed to quantitatively and qualitatively evaluate multiple alternatives and select the most appropriate cross docking system for their business.

## 3.1 Types of cross docking systems

There are many ways to classify cross docking systems. In this book, the methods of cross docking differ based on the *point of consolidation* and the *point of allocation*. The *point of consolidation* refers to *where* in the supply chain the orders for the same customer are physically assembled in preparation for shipment. The *point of allocation* refers to *when* a determination of final product destinations can be made. There are two kinds of final product destination: *pre-allocated* and *post-allocated*. If product is pre-allocated, the final product destination is determined *before* the products are shipped to the CDO's facility. In post-allocation, final destinations are determined only *after* the products are shipped from the supplier.[1] See Figure 31 for a graphic comparison of the different methods.

### Pre-Allocated Supplier Consolidation

This is preferred among cross dock operators because it requires the least handling within the CDO's facility. This method of cross docking is only possible if the following conditions exist:

CHAPTER 3   Phase 2: Planning and Designing a Cross Docking System

- Product can be allocated to specific customers before the supplier ships it.
- The supplier is capable of building and shipping customer-specific pallets based on their specifications. For manufacturing, the supplier must be capable of creating unit loads for specific production runs.

**REAL WORLD EXAMPLE**

### Lever Bros. Supports Cross Docking

Lever Bros., New York has been offering its customers custom pallets for many years. Three or four different brands are used to build specific assortments based on factors like store size and volume. These pallets are marketed as floor-ready displays primarily to support specific promotions. In addition, the company keeps stock pallet display modules on hand all year. These assortments feature mixed SKUs on trays and have their own UPC codes. Any customer can purchase them at any time. Because they are display-ready, these items are shipped from Lever's facilities to its customers' facilities where the pallet is cross docked to the specific store. The company acknowledges that additional costs are incurred to provide these assortments. Thus, it carefully monitors the creation of these pallets and uses activity-based costing (ABC) to determine how many cartons need to be sold in order to justify the expense. Lever reports such a significant increase in sales with its pallet assortments that it is investing in robotic equipment at one of its distribution centers to reduce the costs for building these pallets. (Source: Casper, Carol, "Flow-through: Mirage or Reality," *Food Logistics*, October/November, 1997, p.52.)

**FIGURE 31**

**Comparing Cross Docking Systems**

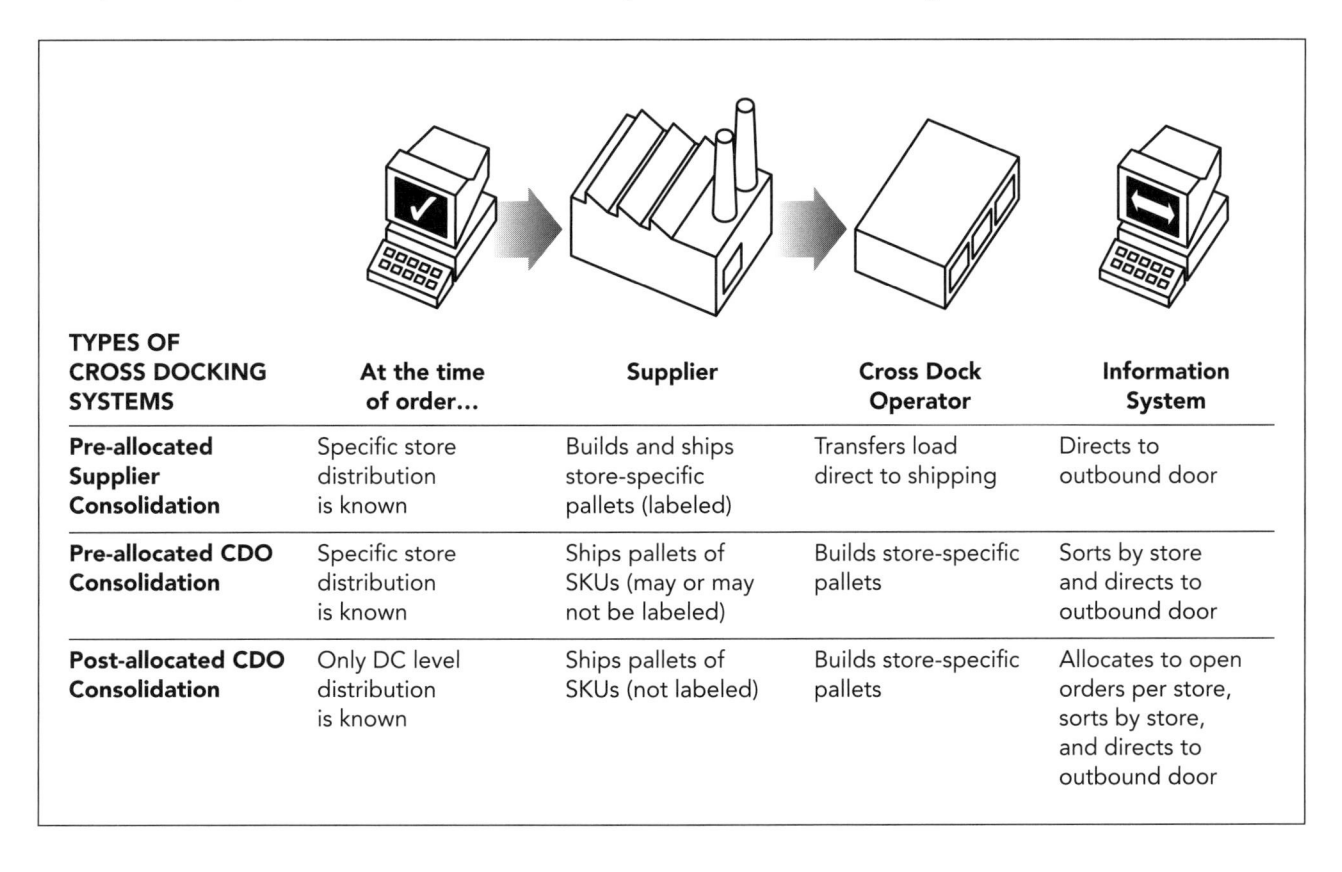

| TYPES OF CROSS DOCKING SYSTEMS | At the time of order... | Supplier | Cross Dock Operator | Information System |
|---|---|---|---|---|
| Pre-allocated Supplier Consolidation | Specific store distribution is known | Builds and ships store-specific pallets (labeled) | Transfers load direct to shipping | Directs to outbound door |
| Pre-allocated CDO Consolidation | Specific store distribution is known | Ships pallets of SKUs (may or may not be labeled) | Builds store-specific pallets | Sorts by store and directs to outbound door |
| Post-allocated CDO Consolidation | Only DC level distribution is known | Ships pallets of SKUs (not labeled) | Builds store-specific pallets | Allocates to open orders per store, sorts by store, and directs to outbound door |

## How It Works In the Supply Chain

1. The supplier receives a purchase order from a customer with instructions on how product should be distributed to each store or assembly plant.

2. The supplier builds a unit load, typically on a pallet, of different SKUs for a specific store or plant.

3. The supplier applies labels on the load, which represents all products on the pallet and, in better systems, includes the end destination.

4. The supplier ships the load to arrive on the same day that the stores are to be replenished by the CDO or that the plants are to use the component parts from the CDO for production.

5. At the CDO's facility, the load is identified, verified against a receiving invoice, and transported to the shipping area.

6. Two things may happen to the unit load. It may be transferred directly to an outbound door, designated for a specific store, plant, or geographic area, and into a waiting trailer, or it may be staged awaiting the arrival of the outbound truck for consolidation with other unit loads for the same customer or geographic area.

7. The unit load is intended solely for one end destination. It is never broken up into individual cartons or pieces.

**FIGURE 32**

**Automated Pallet Load Cross Docking System**

ACTIV System, ICA Partihandel Ab; Helsingborg, Sweden. Commissioned in 1992. (Courtesy of Retrotech, Inc.; 610 Fishers Run; P.O. Box 586; Fishers NY 14453-0586; 716-924-6333 or www.retrotech.com)

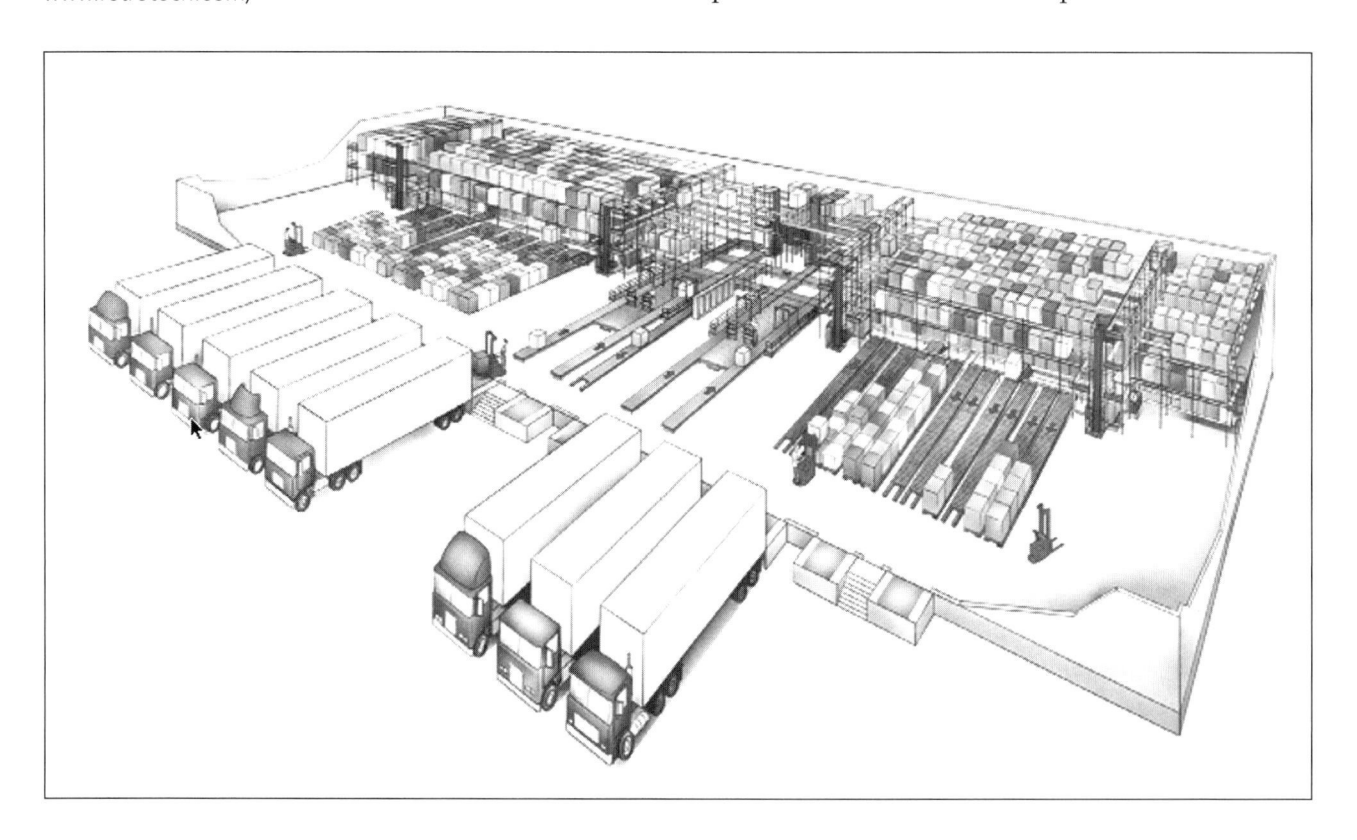

# CHAPTER 3  Phase 2: Planning and Designing a Cross Docking System

8. The unit load is shipped to the store or plant where the contents are received and verified.

## Sample Products That Use This System

- High cube, high demand SKUs for the biggest stores of a retailer
- Floor-ready merchandise or displays pre-assembled by the supplier for a specific store
- Pre-assembled specific store orders from multiple suppliers from another facility
- Pre-assembled component parts of a scheduled production run for a specific plant
- Initial launches of a new product pre-allocated to specific stores
- Promotional products
- Seasonal products (Halloween candy, Christmas decorations, etc.)

## Operation and Equipment Description

- *The manual approach.* This type of cross docking can succeed with the least complex handling equipment consisting of pallet jacks and lift trucks. Counterbalanced lift trucks can move palletized loads from receiving to shipping at a rate of 10-15 pallets per truck per hour.[2] For products that are difficult to stack, stacking frames, drive-in, or drive-thru racks may be used to stage pallets in limited space.

- *The mechanized approach.* Some companies, many abroad, adopt a more mechanized approach to handling pallet loads. Figure 32 illustrates such a system.[3] In this layout example, each unit load arriving at the input conveyor is assigned a pallet ID with product and packaging information. SKU data is matched with the data from the WMS and a barcode label is printed with the necessary information including the final product destination. In some cases, the supplier has already attached the pallet ID to the load before it enters the CDO's facility. The pallet is conveyed to racks where it is sorted in reverse truck loading sequence before being brought out to the outbound dock area just minutes before loading into outbound trailers. Pallet conveyors then move the pallets to the end of the staging lane to simplify and speed the truck loading process. Conveyors may be used to automatically load and unload pallets from trailers as shown in Figure 33.

**FIGURE 33**

**Automatic Pallet Unloading Conveyor**

(Courtesy Gross & Associates)

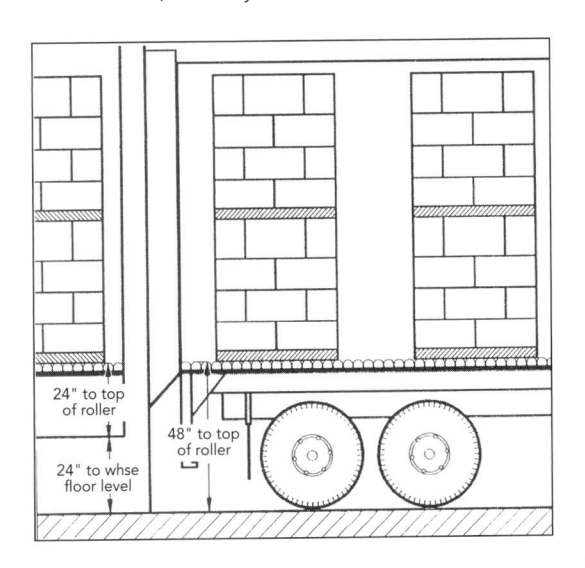

The system shown here handles more than 2,000 pallets through the system per 12 hour day with only 1,500 pallet positions. The peak capacity of the facility is 300 inputs/360 outputs per hour for a combined in/out peak rate of 500 pallets per hour.[4] With 10 lift trucks in operation, a manual system would have only a peak capacity of 150 pallets per hour.

**Information Systems Description**

- *The manual approach.* With information, cross docking full pallets is not a difficult process. Some companies do just that with little or no use of the latest information technology. In one example, a third party provider consolidates pre-picked orders from multiple suppliers at its cross dock center with neither bar codes nor radio frequency devices. Suppliers are notified of delivery schedules to customers' stores via fax or phone. Pre-picked orders arrive on pallets at the cross dock center with human-readable labels attached to indicate the pallets' end destination. CDO personnel read the information on the pallets and transfer the loads to the appropriate staging lane for a particular store. Notices of impending shipments are sent via fax to the stores and any delays are called in by phone. Although largely paper driven, this particular facility is capable of receiving and shipping all orders in fewer than 18 hours.

- *The automated approach.* Increases in throughput volumes and the number of store allocations will accentuate the need for more sophisticated, "high-tech" information systems even with conventional equipment such as lift trucks and pallet jacks. These systems not only increase efficiency but give real time, predominantly error-free and paper-free information to the user. Examples of these technologies include EDI with ASNs, bar coding, and RF communications. With bar coding and RF, for example, workers or omni-directional scanners (for automated systems) can scan an item bar code on the pallet and feed the information into the WMS software. The software consults its database to determine the appropriate dock door for that item and transmits this information back to the worker or to conveyor controls—all in a matter of seconds. The following steps illustrate a sample supply chain process flow of an automated information system for this type of cross docking in a retail environment (also see Figure 34).

    1. POs are transmitted from customer to supplier via EDI with store-specific allocation.

    2. Supplier acknowledges the PO and confirms the shipping date also through EDI.

CHAPTER 3   Phase 2: Planning and Designing a Cross Docking System

3. Supplier builds the store-specific pallet, labels the pallet with a master bar code representing all cartons on the pallet, ships it, and sends an ASN to the CDO.

4. Cross dock software at the CDO's facility begins to plan outbound loads before the products arrive at the facility.

5. When the trailer does arrive, an inbound receiving employee scans the pallet bar code, using either RF terminals on lift trucks or wall-mounted terminals.

6. The software then directs each shipment to the proper staging lane or directly to an outbound door.

7. Bar codes on the pallets are scanned again as freight is loaded onto the outbound truck. This is done not only to confirm and validate status, but also for trailer manifesting.

8. When the products are on their way, an ASN is sent to each store informing it of the incoming arrival.

9. The store receives merchandise and sends confirmation back to the CDO.

**FIGURE 34**

Information Cycle for Pre-allocated Supplier Consolidation

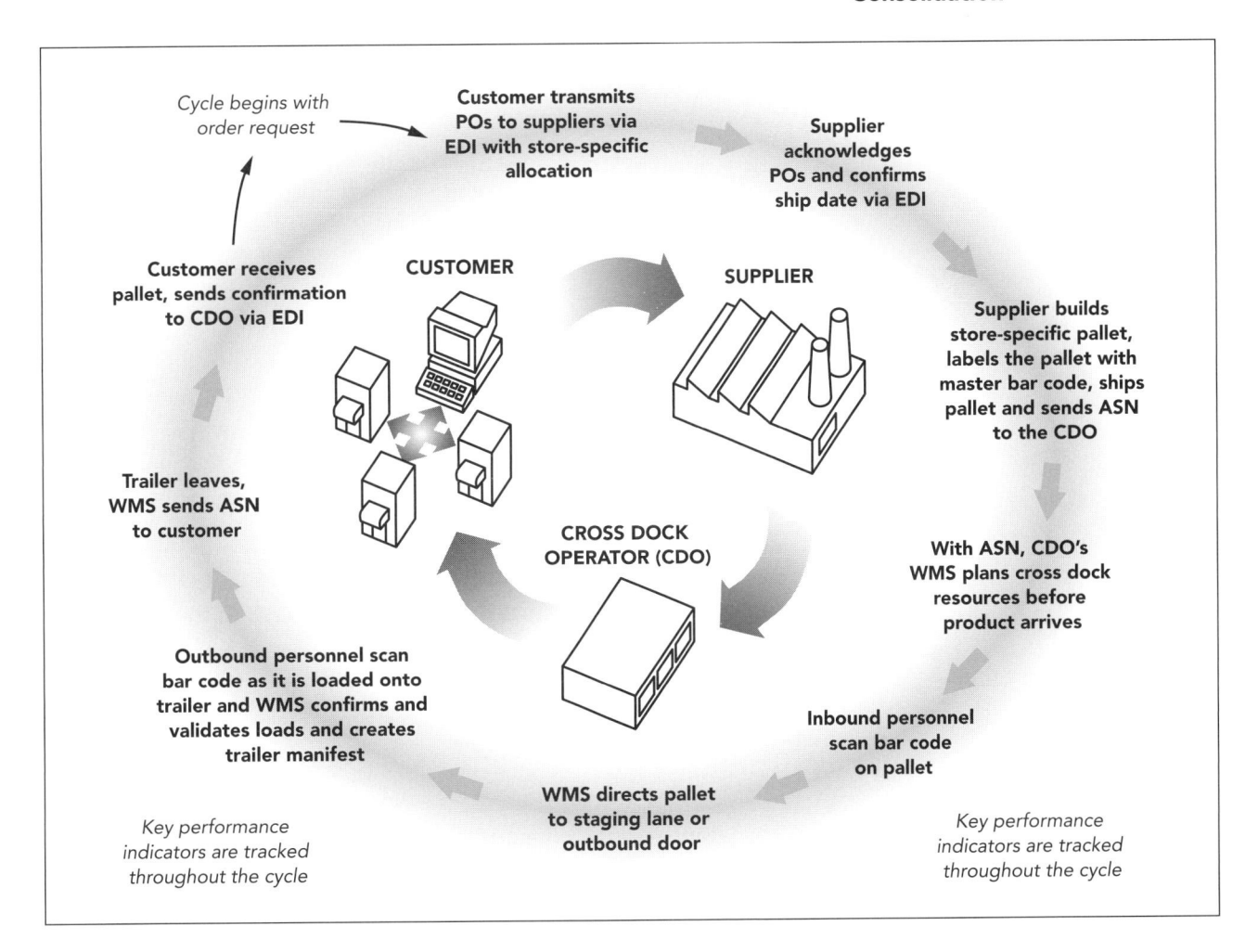

The cycle begins again with the next order request. Throughout the entire process, information systems collect key performance indicators for evaluating the effectiveness of the operation.

## Facility Description

In a full pallet load operation, the CDO's facility should resemble a carrier terminal—usually a long, narrow rectangular building with dock doors along the length of the building across from each other (see Figure 35). When little or no staging is involved and the movement of unit loads is continuous from one truck to another, then the facility can be as narrow as possible. For this to happen, careful, controlled scheduling is critical to coordinate receipts and shipments. This procedure works best when there is a trailer at the shipping dock for each customer's store or plant serviced by the facility—a concept known as *door-per-store*.

Staging space may be preferred for flexibility. Poor scheduling of dock doors, inadequate number of dock doors, and inaccurate information regarding the arrival and quantity of products to be loaded/unloaded increase the requirements for staging space.

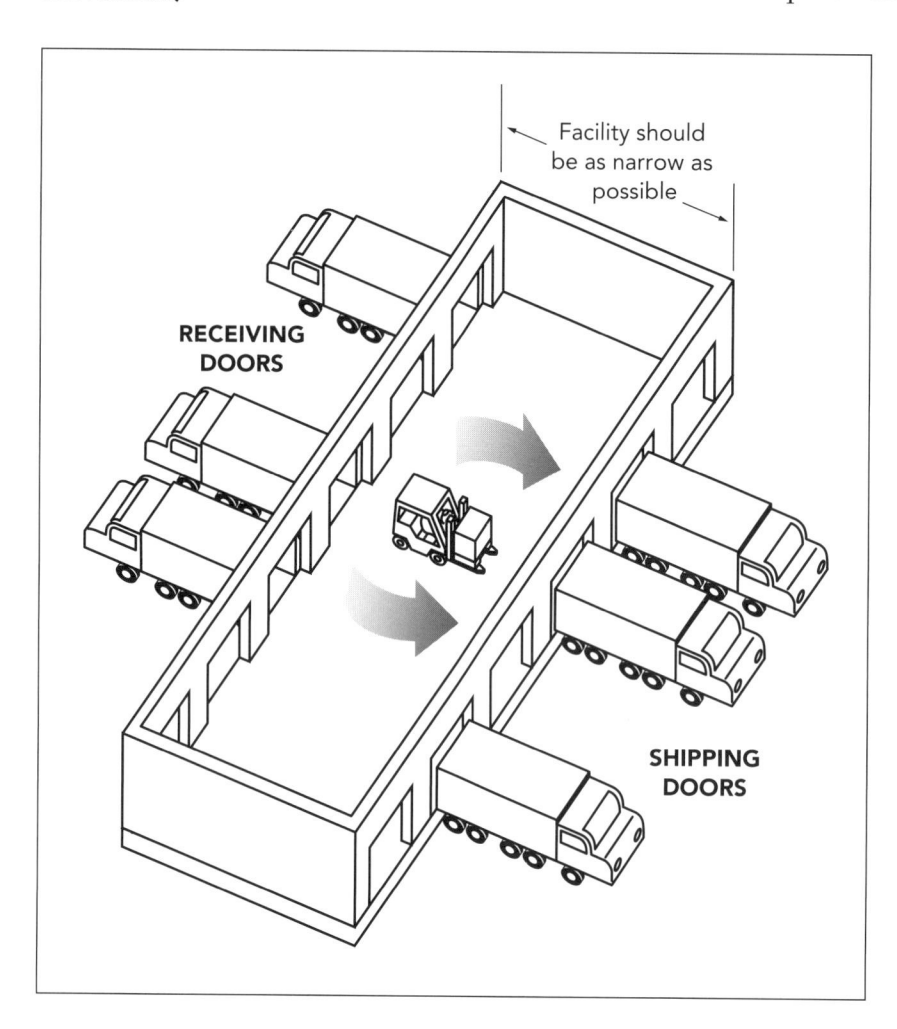

In addition to staging inside the facility, staging may be done at the dock doors. Staging trailers before receiving and staging empty trailers for shipping allows for more flexibility since transportation times vary.[5] A yard tractor and a driver should be available to move trailers as needed.

Of all the types of cross docking, this is the easiest to retrofit in an existing building because complex material handling equipment is not needed. Ideally, the flow from receiving to shipping should be smooth and quick. Storage racks may have to be removed to increase dock space. If the ceiling height in the dock area is low—maybe because of offices on a mezzanine—installing pallet racks for staging will be limited. The correct quantity and physical functionality of

**FIGURE 35**

**Ideal Facility for Pure Supplier Consolidation (Full Pallet Movement)**

existing dock doors with appropriate safety mechanisms are critical to this type of operation.

## Points to Consider

- *Significant reduction in warehousing activities.* This is the simplest and most popular type of cross docking method. It offers significant reduction in the CDO's warehousing activities as product literally flows in and out of the facility with the least amount of handling and with little or no staging.

- *No inventory.* Each incoming unit load has a pre-established destination outside of the CDO's facility, thus storage space is not necessary.

- *Heavy dependence on suppliers.* The assembly of pallets relies heavily on suppliers building custom pallets based on their customers' specifications. For maximum restocking efficiency at the store level, retailers may want their pallets built to correspond to store layouts. Cross dock programs have sometimes failed because products do not arrive at the stores in correct family groupings, and incremental store labor is spent putting them away. In some cases, the size of the unit load shipped by the manufacturer does not conform to the CDO's rack configuration or conveyor specifications. The receivers may have to sort, segregate and restack the load. Unloading time then increases from an average of 30 minutes to more than an hour and a half.[6] However, with cross docking practices being pushed by the larger retailers, an increasing number of suppliers are building customized pallets based on factors like store size and volume and their customers' unique specifications.

- *Limited to a chosen few.* In retail, the cross docking of single SKU unit loads is most economical for ultra fast-moving, high cube, and/or low value items for delivery to a customer's biggest stores. Not many SKUs are meant for such high density, rapid movement, since even a retailer's biggest stores do not always have adequate backroom storage for a pallet of only one SKU.[7] Custom or multi-SKU pallets address this issue by allowing multiple SKUs in the same amount of space. In manufacturing, unless a full pallet can be inducted immediately into a production run upon receipt, the same concern also applies. Pre-assembly staging area for component parts is very limited.

- *Works with cross docking of slow movers and other specialty products.* Some companies outsource low velocity SKUs to specialty distributors or third party providers and thereby eliminate thousands of items from inventory.[8] Other retailers prefer to retain control by pulling out slow movers from their own regional DCs and consolidating them in a separate DC. Retail stores then transmit orders for slow-moving SKUs to this slow moving/specialty facility where the orders are picked, packed, and shipped to the CDO's facility where it is cross docked to the appropriate store. Either way, the results are net savings for the retailer.

## Pre-Allocated Cross Dock Operator Consolidation

With the first method, the supplier is required to assemble SKUs on a pallet. In this next method, the responsibility of segregating products by plant or store is placed on the cross dock operator. Sorting at the carton level increases the complexity of the process for the DCO. Product is still pre-allocated in that each carton has a known destination before it enters the CDO's facility. The allocation of SKUs is also predetermined.

## How It Works In the Supply Chain

1. The supplier receives purchase orders from the customer that indicate aggregate requirements at the facility or distribution center level. The customer may or may not include instructions on how products should be labeled for the individual plant or specific store.

2. If instructions are included, the supplier attaches labels to each carton before consolidating by SKU. If no instructions are included, the supplier simply consolidates all products by SKU before shipping them to the CDO's facility.

3. The supplier ships the load to arrive on the same day that the customers are to be replenished by the CDO.

4. Before the SKUs arrive at the CDO's DC, the supplier sends the CDO a detailed report of the contents of the impending shipment. (SKUs, number of cartons per SKU, ship date, etc.)

5. Since specific allocations are made before the time of receipt at the CDO's DC, each item has a known destination after it arrives at the CDO's facility.

6. If labeled by the supplier, product will enter the CDO's facility ready to flow through the system. If not, the CDO applies the appropriate labels.

7. Product is then transported either by conventional lift trucks or conveyors to a sortation area where cartons bound for the same customer are consolidated and palletized.

8. The consolidated unit load is transferred either directly to an outbound truck or to a staging area for later loading.

CHAPTER 3  *Phase 2: Planning and Designing a Cross Docking System*     77

9. The unit load is shipped to the plant or store where the contents are received and verified.

## Sample Products That Use This System

- All of the sample products from the previous method
- Mixed pallets or less-than-pallet store order quantity
- Any pre-labeled cartons that indicate end destination customer

## Operation and Equipment Description

- *The manual approach.* This method of cross docking can be accomplished with pallet jacks and lift trucks. There are many variations of manual sort and load cross docking operations. One example is called the *sort-to-pallet* method. Unit loads are unloaded from trailers by lift trucks and transported to a staging area for sorting. As part of this unloading process, a distribution list for each SKU is generated to indicate the quantity to be distributed to each customer. This list should be generated by SKU, regardless of how many different SKUs are received on a trailer. In addition, this list should be sorted in the same sequence that the lanes are assigned to customers in the cross dock aisle to avoid backtracking by warehouse personnel.[9] Using a pallet jack, a warehouse person will take one pallet at a time, and travel along the cross dock aisle, placing the appropriate number of cartons on the pallet in front of a lane designated for each customer. The process will continue as each SKU is received. When the pallet on the floor is filled, a customer specific label is attached to the load before it is placed in the next available position in a staging lane. Another empty pallet is then placed on the floor. At the other end of the lane, lift trucks are used to pull completed, customer specific pallets from the lane and loaded into trailers.

- *The mechanized approach.* The mechanized method of consolidating orders uses conveyors, an automatic sortation system, and leading-edge information technology. (Figure 36 illustrates a mechanized cross

**FIGURE 36**

**Mechanized Cross Dock System Layout**

(Courtesy of Rapistan Systems; 507 Plymouth Avenue NE; Grand Rapids, MI 49505-6098; 616-913-6525)

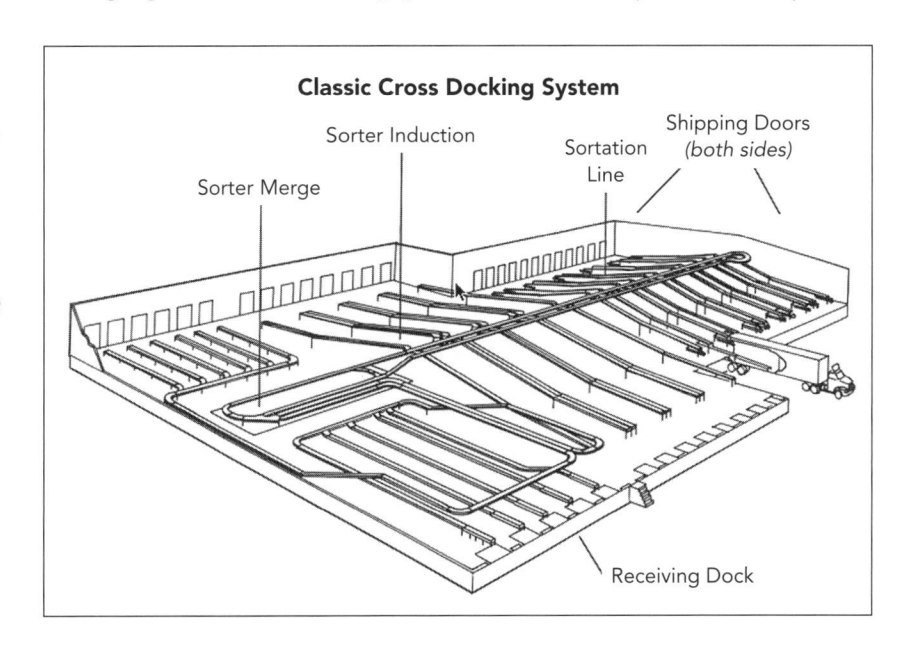

## FIGURE 37

**Conveyor Equipment for Receiving and Sorting Cross Docked Products**

*Top left:* At receiving, cases can be labeled immediately—while they are being unloaded from trailers. *Top right:* Cases can be automatically merged while they are conveyed to the sorter. *Bottom:* A positive shoe sorter can provide a high rate of computer-controlled sortation to either side, or it can sort bi-directionally.

docking layout.[10]) At the receiving dock, incoming shipments are unloaded by workers who break down palletized unit loads and place the cartons on extendable conveyors. (See Figure 37 for receiving equipment examples.) Large, bulky, non-conveyable loads bypass the conveyor system and are transported using conventional lift trucks and pallet jacks. If labels were not applied by suppliers prior to being received at the CDO's facility, then a bar coded shipping label is applied to each carton as it is removed from the trailer. A network of conveyors transports each carton to one or more omni-directional laser scanners that read the bar codes on the attached labels (see Figure 38). The scan is transmitted to a Programmable Logic Controller (PLC), which provides the intelligence to direct each carton either to storage or to the appropriate outbound shipping lane. The PLC then communicates with the WMS in order to tally inventory, update purchase orders, and initiate payment of invoices. Cartons are then diverted and accumulated in their designated conveyor lanes close to the shipping doors. Workers palletize product into unit loads or, using extendable conveyors, load the cartons into the outbound trailer. (See Figure 39 for an illustration of this concept.) Scissor lifts may be used at palletizing stations to maintain proper working heights and to reduce unnecessary bending and lifting. Companies cross docking more than 70 to 90 percent of their merchandise rely on automated sortation systems to achieve throughput rates of over 40,000 cartons per eight-hour shift or

CHAPTER 3  Phase 2: Planning and Designing a Cross Docking System          79

100 to 200 cartons per minute. In these systems, it takes a total of 20 minutes for an incoming carton to be unloaded, bar coded, transported, sorted, and loaded into an outbound trailer.[11]

## Information Systems Description

- *The manual approach.* A "low-tech" system for exchanging information for this type of cross docking can be employed for low to medium throughput volumes with allocation to relatively few customers. For higher volumes, more sophisticated systems are necessary. Pre-labeled cartons are easier to process

**FIGURE 38**

### Equipment for Sorting Cross Docked Products

*Top left:* Bar-code information on labels can be read automatically as cases pass under a high-speed laser scanner. *Top right:* At lower speeds, an operator can read bar-codes with a hand-held scanner, or manually enter information at a console. *Bottom:* A steerable roller sorter rapidly diverts cases to takeaway lines leading to shipping doors.

**FIGURE 39**

### Conveyor Equipment for Loading Cross Docked Products

*Below left:* A skewed pop-up wheel sorter can be used to provide computer-controlled sortation at mid-range speeds. *Below right:* Declining belt and/or gravity roller conveyor units can be used to convey cases from a mezzanine-level sorter to waiting trailers.

(Photos courtesy of Rapistan Systems; 507 Plymouth Avenue NE; Grand Rapids, MI 49505-6098; 616-913-6525)

manually, because the warehouse worker can read the necessary information off the label and consolidate the merchandise with other cartons destined for the same customer. For the manual approach illustrated above, the Warehouse Management System will print a distribution list for each single SKU palletized load to determine the quantity allocated for each customer. To do this, receiving personnel must enter inbound shipment information upon receipt. After products are consolidated, a checking station is required to verify the manual sort.

- *The automated approach.* As stated earlier, a mechanized approach to this type of cross docking relies heavily on leading-edge information systems technology. The most prominent technology for this application includes bar coding, WMS communication with the conveyor controls, and scanning devices for identifying and sorting product. The following steps illustrate a sample information systems cycle flow for an automated approach in a retail environment (also see Figure 40).

**FIGURE 40**

Information Cycle for Pre-allocated CDO Consolidation

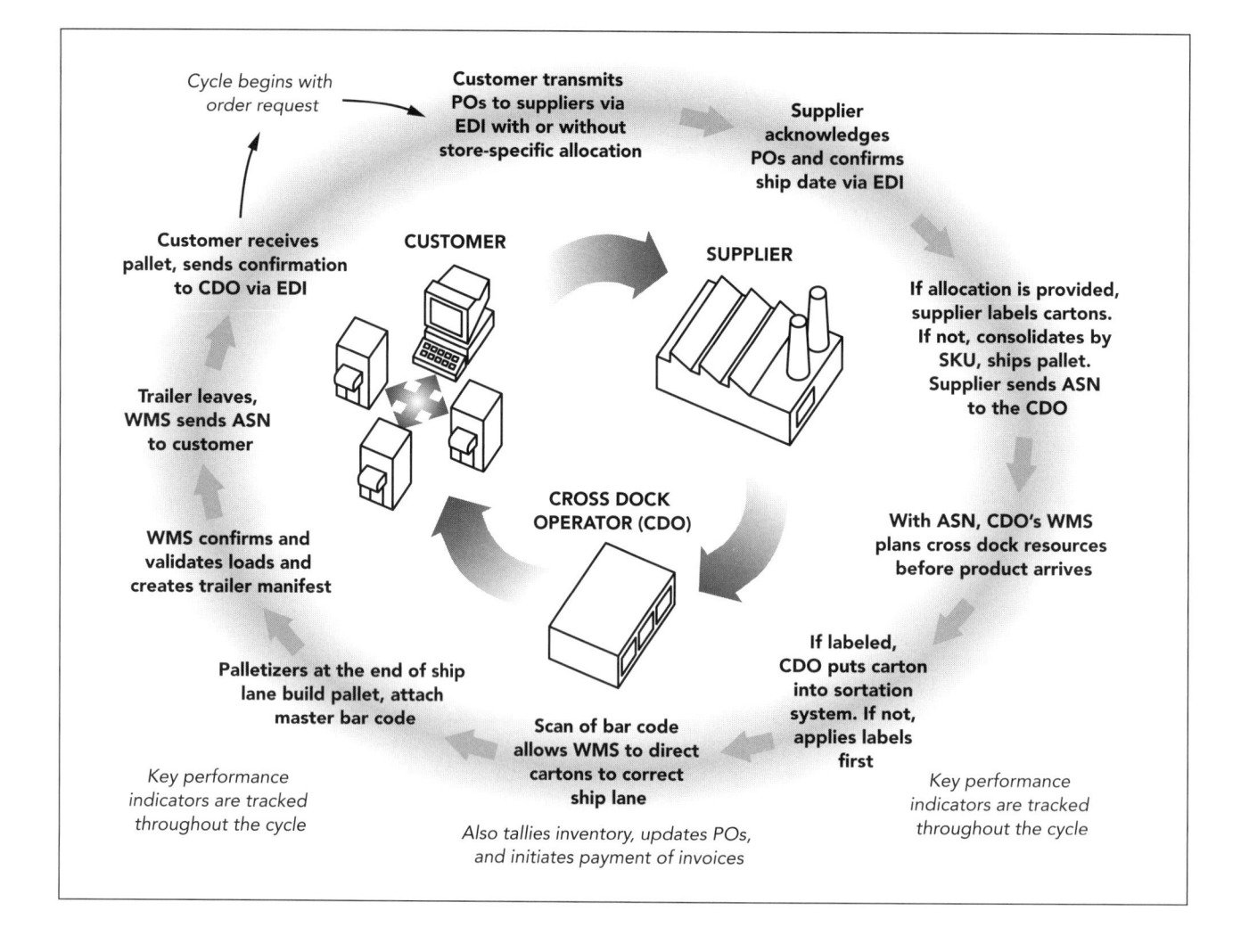

*Cycle begins with order request*

**Customer transmits POs to suppliers via EDI with or without store-specific allocation**

**Supplier acknowledges POs and confirms ship date via EDI**

**CUSTOMER**

**SUPPLIER**

**If allocation is provided, supplier labels cartons. If not, consolidates by SKU, ships pallet. Supplier sends ASN to the CDO**

**Customer receives pallet, sends confirmation to CDO via EDI**

**Trailer leaves, WMS sends ASN to customer**

**CROSS DOCK OPERATOR (CDO)**

**With ASN, CDO's WMS plans cross dock resources before product arrives**

**WMS confirms and validates loads and creates trailer manifest**

**Palletizers at the end of ship lane build pallet, attach master bar code**

**Scan of bar code allows WMS to direct cartons to correct ship lane**

**If labeled, CDO puts carton into sortation system. If not, applies labels first**

*Key performance indicators are tracked throughout the cycle*

Also tallies inventory, updates POs, and initiates payment of invoices

*Key performance indicators are tracked throughout the cycle*

1. POs are transmitted from customer to supplier via EDI at the DC or facility level with or without store-specific instructions for labeling cartons.

2. Supplier acknowledges POs and confirms the shipping date through EDI.

3. If instructions are provided, the supplier labels the cartons with specific labels and consolidates the cartons by SKU. If not, the supplier ships the required quantity of a SKU without any labeling. Once shipped, an ASN is sent to the CDO.

4. Cross dock software at the CDO's facility begins to plan outbound loads before the product arrives at the facility.

5. When the trailer arrives, inbound receiving personnel scan a bar coded trailer license plate that verifies the contents of the incoming shipment against the ASN from the supplier.

6. If cartons are pre-labeled, a receiving person unloads the cartons onto conveyors that transport them to omni-directional laser scanners. If cartons are not labeled, the receiving person prints out the appropriate bar coded labels to attach to the incoming cartons. (This can be automatically generated in the previous step. In some cases, an automatic labeling machine may be used.)

7. Scanning the bar codes on the cartons enables the WMS to direct them to the appropriate shipping lane or storage location, tally inventory, update POs, and initiate payment of invoices. The WMS transmits the necessary information to the conveyor controls so that cartons can be diverted to the appropriate outbound accumulation conveyor lane.

8. Palletizers at the end of this lane build pallets specifically for a particular store. A master bar code is attached to the pallet representing all the cartons on that pallet.

9. Bar codes on the pallet are scanned again as they are loaded onto the outbound truck. This is done to confirm and validate status and for input into the trailer manifest.

10. After product is on its way, an ASN is sent to each store informing it of incoming merchandise.

11. Each store receives the merchandise and sends confirmation via EDI back to the CDO.

The cycle begins again with the next order. Throughout the process, key performance indicators are collected for evaluating the effectiveness of the operation.

## Facility Description

Sortation requirements will dictate the space and type of facility needed for this method of cross docking. At a minimum, sortation operations will require a dedicated area close to the dock doors for breaking down and palletizing customer specific unit loads. High throughput volumes and increased store allocations require automated sortation systems and a specialized facility to house these systems.

The correct number and position of dock doors is also critical to an efficient cross docking operation. Dropping a trailer (as in a door-per-store scenario) may eliminate staging space because product is immediately loaded, but it ties up the dock door while waiting cube-out volume. This is a space-saving strategy for high throughput operations. Determining the correct number of doors to use takes these strategies into consideration.

Additional staging space is needed to stage product before and after sorting. These staging areas need to be adjacent to inbound and outbound dock doors.

Retrofitting an existing facility that already has sortation systems in place should not be difficult for this method of cross docking. Dock space usually takes the place of displaced storage. Retrofitting a heavily-racked traditional store-and-ship facility to accommodate a mechanized version of this method of cross docking is not recommended. Installing conveyors and sortation systems is complex in itself and next to impossible to implement without disrupting existing operations. Therefore, a mechanized approach is more suited for a facility built specifically for large-scale sortation and cross docking.

## Points To Consider

- *Works well for less-than-pallet, carton quantity orders per customer or store.* Pioneers in cross docking philosophy suggest that the breakdown of unit loads realizes the bulk of cross docking's potential.[12] Ordering at the carton level makes it possible to cross dock a wider array of SKUs without crowding valuable retail backroom storage and pre-production staging space. Retail stores and assembly plants are in a better position to absorb the merchandise than in the previous method.

- *Additional handling by the CDO required.* This type of cross dock requires that the CDO invests labor and equipment to sort incoming products by customer and store. With a manual approach, sortation can be very labor intensive, and the additional handling can negate the benefits of cross docking. Sortation equipment can replace labor in this scenario, but a substantial capital investment will be required.

CHAPTER 3   Phase 2: Planning and Designing a Cross Docking System

- *No inventory.* Each incoming unit load has a pre-established destination outside of the CDO's facility; therefore storage space is not necessary. Items are totally eliminated from inventory.
- *Partial reliance on suppliers.* There is still reliance on suppliers to provide correct quantities and, when labeling is involved, to attach the correct label to each carton. Some contingencies will be needed to counteract any errors by suppliers.
- *Higher initial investment.* The more automated the system, the higher the initial capital investment.

## Post-Allocated Cross Dock Operator Consolidation

In the previous method, the CDO has determined the product allocation for specific customers' stores or plants before the product is shipped, based on some scheduled production run or forecasts or marketing analysis for the region. In this method, there is no pre-allocation and no external reliance on the supplier. A product's final destination may not be determined until after it is en route to the CDO's facility. In this approach, the ideal product to cross dock has a constant and continuous demand pattern that is easy to predict. In addition, to fully maximize the benefits of such a system, the CDO has information and material handling systems that enable handling the physical product to remain in synch with the flow of information in real time. Once received, cross docked products may be placed in a "hot processing area" and used for immediate order selection. Replenishment of the product is on a just-in-time basis from the supplier—eliminating the need for reserve storage. Product is still sorted and palletized by specific customer or store.

## How It Works In the Supply Chain

1. The supplier receives purchase orders from the customer that indicate aggregate requirements at the facility or distribution center level and the required shipping dates.
2. The supplier consolidates all products by SKU before shipping them to the CDO's facility.
3. The supplier ships the product to arrive shortly before the store or plant is to be replenished.
4. Before the SKUs arrive at the CDO's DC, the supplier sends the CDO a detailed, comprehensive report of the contents of the incoming shipment (SKUs, number of cartons per SKU, shipping date, etc.).
5. Hours before the trailer arrives, the CDO starts allocating the SKUs arriving that day to current orders and calls for replenishment to specific customers scheduled for delivery.

6. Product is received at the CDO's facility and identified as "hot product."
7. Product is then transported to a "hot processing area" near the shipping dock.
8. Warehouse personnel pick, label, and load orders for the customer as required.
9. At the end of the day, remaining cartons, if any, may either be staged for processing the following day or placed in storage.
10. The consolidated unit load is transferred either directly to an outbound truck or to a staging area for later loading.
11. The unit load is shipped to the end destination where the contents are received and verified.

**Sample Products That Use This System**

- Less-than-pallet store order quantities ordered at predictable levels on a replenishment basis
- Staple products or products with a constant, predictable demand
- Raw components for continuous production manufacturers

**Operation and Equipment Description**

- *The manual approach.* A pure, manual approach (no WMS or radio frequency) to this type of cross docking is difficult, if not impossible. With post-allocation, personnel have to communicate with the WMS in real time or very close to it to determine where products should go. However, conveyor equipment and automated sortation systems are not necessary to make this type of cross docking succeed. In fact, the *sort-to-pallet* approach of moving and distributing the cartons of a single SKU pallet through an aisle of customer-specific pallets can be used. For each hot item, the WMS must prioritize and determine which open order should be filled by this item for immediate shipment. The WMS then directs personnel through radio frequency of the appropriate number of cartons for each store. (See the information systems section also.)

    Another manual approach is the *pick-to-pallet* method, in which cross docked or "hot items" are placed in the forward picking line directly from receiving. Using pallet jacks, pickers travel down a forward picking line filling orders in carton quantities for a customer. At the end of the day, any remaining cartons will be processed for the following day. If post-allocated SKUs are cross docked in full pallet quantities,

CHAPTER 3  *Phase 2: Planning and Designing a Cross Docking System*  85

warehouse personnel with radio frequency terminals on their lift trucks or pallet jacks transfer the unit loads to the appropriate customer lane based on WMS instructions.

- *The mechanized approach.* The mechanized method of allocating SKUs and consolidating orders still makes substantial use of conveyors and leading-edge information technology. It essentially uses the same equipment and operation used in the mechanized approach of cross docking pre-allocated products except that a bar coded shipping label has to be applied to each carton before being inducted into the sortation system. When the sortation system does not have enough divert lanes for the number of stores, picking and customer-specific palletization can be done in waves. After a wave, each divert lane is used to accumulate cartons for a different store. Although more complicated, cross docking can still be used in such a system using a *pick-to-belt* method. Cross docked material from receiving is placed directly in a fast-moving lane where workers label and place the appropriate cartons for a wave on a conveyor that feeds into an automatic sortation system.[13] See Figure 41 for a close-up view of a pick-to-belt cross docking operation leading to an automatic sortation system. From the warehouse worker's viewpoint, the flow and handling of products is the same whether it is pre-allocated or post-allocated. It is with the information systems and the software that a major difference can be expected.

**FIGURE 41**

**Product Flow for Pick-to-Belt with Automatic Sortation System**

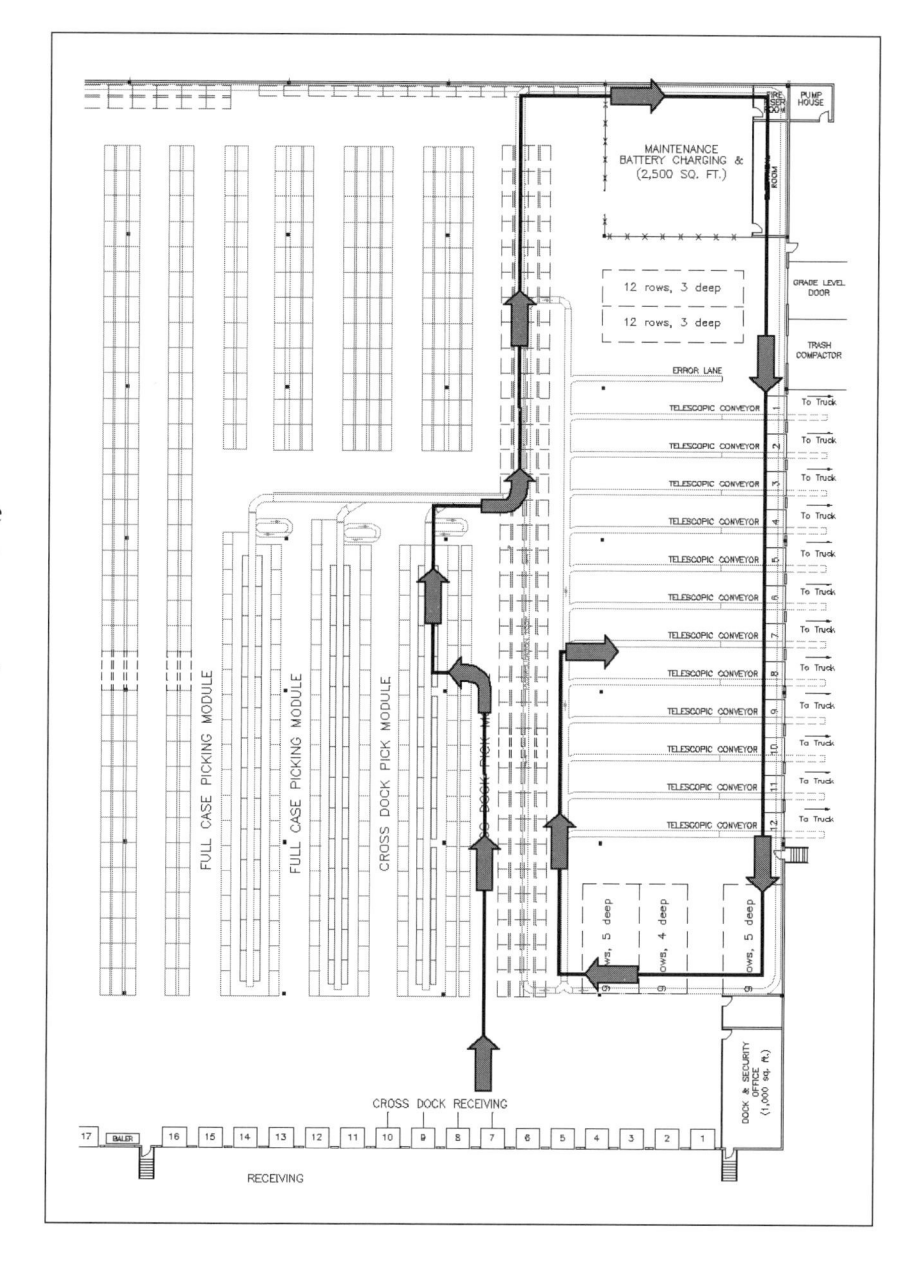

### Information Systems Description

- *The manual approach.* A "low-tech" system for exchanging information for this type of cross docking is not recommended. At a minimum, the CDO must have a WMS capable of allocating product when it arrives so that distribution lists can be printed for warehouse personnel to use for sorting by stores in the *sort-to-pallet* method. For *pick-to-pallet* operations, paper-driven systems use picking lists for each customer or store. The cross docked items would be located at a permanent picking position and pickers would pick the appropriate number of cartons for each store. However, the numerous checks required by the manual exchange of information can slow down what should be a fast operation.

- *The automated approach.* Post-allocated cross docking relies heavily on a WMS's ability to distribute incoming products to open orders before they arrive. In addition, the WMS must be able to communicate this allocation in real time to sortation equipment and personnel so that merchandise can be labeled and transferred to the appropriate sortation lane on the shipping dock for customer-specific order consolidation. Bar code scanning, radio frequency, and accurate allocation algorithms all help to improve the efficiency of handling the information so as to keep pace with the physical handling of the product. The following steps illustrate the typical flow of information with automated systems in a retail environment (see also Figure 42).

  1. POs are transmitted from customer to supplier via EDI at the DC or facility level.

  2. Supplier acknowledges POs and confirms the shipping date, also through EDI.

  3. Supplier ships the required quantity of an SKU. After shipping, an ASN is sent to the CDO via EDI.

  4. Hours before the shipment arrives, WMS starts allocating incoming SKUs to open orders and calls for replenishment:

     a) If the quantity received is *less than* the quantity required, the difference can be made up by a small inventory stored in the facility.

     b) If the quantity available is still *less than* the quantity required, the quantity allocated to stores has to be reallocated based on an agreed-upon formula for reducing the quantities to each store.

     c) If the quantity received is *more than* the quantity required, the difference may either be moved to storage or staged at the same location for processing the next day. In either case, WMS must add the "left over" quantities to the quantities available for allocation on the following day.

5. When the trailer arrives, an inbound receiving person scans a bar coded trailer license plate that verifies the contents of the incoming shipment against the ASN from the supplier.

6. Full pallet loads, if any, are directed via radio frequency to the appropriate staging lane for a specific store, and they are labeled and staged for shipping. Cartons are directed to a labeling area where a person prints out the appropriate bar coded labels and attaches them to the incoming cartons. (In some cases, an automatic labeling machine is used.)

7. Scanning the bar codes allows the WMS to direct the cartons to the appropriate shipping lane or storage location, tally inventory, update POs, and initiate invoice payment. The WMS transmits the necessary information to the conveyor controls so the cartons are diverted to the appropriate outbound accumulation conveyor lanes.

**FIGURE 42**

**Information Cycle for Post-allocated CDO Consolidation**

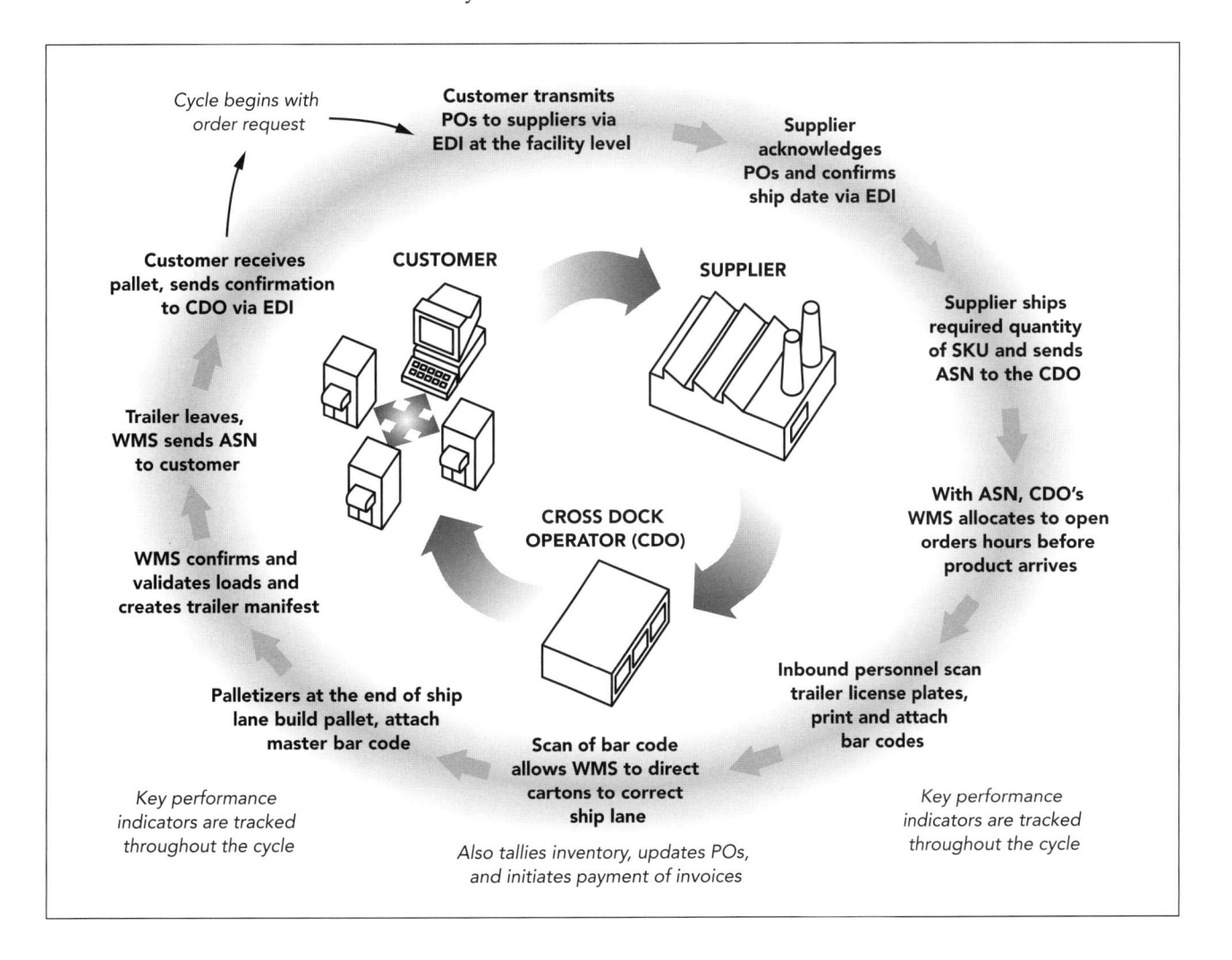

8. Palletizers at the end of a lane build pallets specifically for a particular store. A master bar code representing all the cartons is attached to the pallet.

9. Shipping bar codes on the pallet are scanned again as freight is loaded onto the outbound truck. This is done to confirm and validate status and for input into the manifest.

10. As soon as product is on its way, an ASN is sent to the particular store informing it of incoming merchandise.

11. The store receives the merchandise and sends confirmation via EDI back to the CDO.

The cycle begins again with the next order request. Throughout the entire process, key performance indicators are collected for evaluating the effectiveness of the operation.

## Facility Description

In this scenario, the flexibility of the facility design is a key criterion for cross docking success. Because of demand variations that exist in this fast-flow, pull environment, an inventory may be maintained for SKUs cross docked using this method. The facility should adopt a material flow pattern that facilitates the transport of products from receiving to shipping, from receiving to storage, or from storage to shipping (see Figure 43).[14] SKUs may simply stay in the "hot processing" area for shipping during the following day. Whether they are moved to storage or kept in the picking area, space should be allocated to accommodate these cross docked SKUs.

Retrofitting a traditional storage facility for this method of cross docking may be as simple as dedicating a portion of the forward pick positions to "hot items." Ideally, this location should be close to the receiving doors and along the picking path or, if all products are to be sorted by an automated sortation system, adjacent to an induction lane that is close to the receiving docks. Retrofitting

**FIGURE 43**

**Suggested Flow Pattern for Combining Cross Dock with Storage**

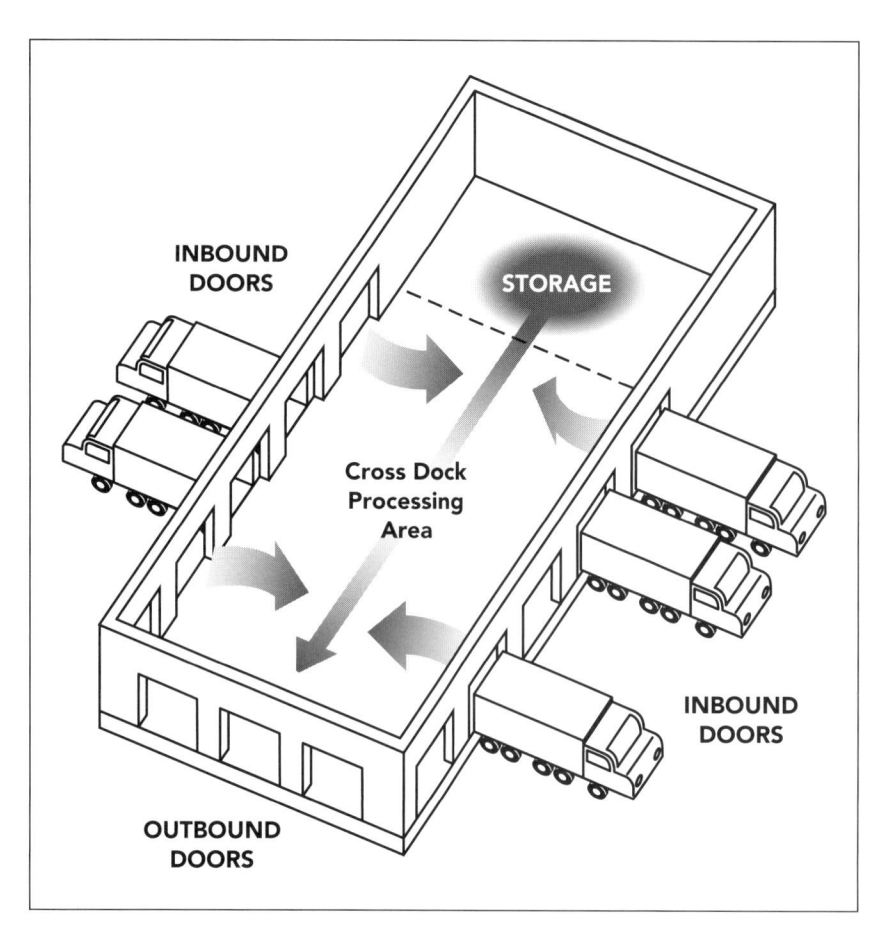

a heavily racked traditional store-and-ship facility to accommodate a mechanized version of this method is not recommended. A fully mechanized approach is more suited for a new facility built specifically for large-scale sortation and cross docking.

**Points To Consider**

- *Little or no reliance on the supplier.* The CDO does not have to depend on the supplier to correctly label and consolidate customer specific orders. However, the supplier is still expected to deliver the correct quantity of the correct product at the specified time.

- *Additional handling for the CDO.* Including a picking step increases handling requirements for the CDO and may negate the benefits of cross docking. However, mechanization and real-time information systems can eliminate picking. In receiving, the WMS prints the appropriate labels for a cross docked SKU and a receiving person (or an automatic labeler) attaches the label to the carton. Then the cartons are inducted into a sortation system that automatically sorts and consolidates orders for the same store. Careful economic analysis must be performed before such a system is implemented. Real-time inventory control and the operation of complex material handling equipment can cost a company $250,000 to $2 million for the software package and its integration with existing computer systems.[15]

- *Inventory is substantially reduced, but may not be totally eliminated.* Because cartons are not pre-allocated and because unforeseen circumstances can make demand unpredictable, carrying a small inventory of an item is recommended and can be very helpful for reducing out-of-stock conditions in stores and plants. Companies utilizing this approach have gone from six weeks of inventory to a few days. Be wary of maintaining too much inventory because it conflicts with cross docking objectives.

- *Works with processing open replenishment orders from the previous day for immediate delivery.* Before the trailer arrives at the CDO's facility, the WMS can match incoming SKUs to open replenishment orders based on POS sales data from the end of the previous day. Any remaining cartons may be placed in storage or staged for processing the following day. Inventory at the customer and at the CDO's facility is kept at a minimum.

- *Faster cycle times.* If product is distributed to current open orders at the time of receipt in the DC, replenishment cycle times will decrease. At the end of the day, depending on delivery schedules, replenishments are literally on their way to the stores. The cross docked product is received and shipped on the same day over one or two shifts.

- *Works well for less-than-pallet carton quantity orders per SKU per customer or store.* Customers in a pull system benefit by carrying only a small inventory in stores or plants.

- *Higher initial investment.* Post-allocated cross docking has a higher initial investment than pre-allocated cross docking. More sophisticated software is usually required for allocating incoming product to open orders. Carrying inventory into the operation also adds costs.

**Third Party Cross Docking**

This last type of cross docking incorporates any one of the methods. It is unique in that a *third party provider* assumes the role of cross dock operator. Third party providers are able to pass on to customers economies of scale, allow sharing of resources and expertise, and provide better vehicle use than their logistics counterparts in private industry. In some specific markets, a manufacturer or retailer with its own fleet is unlikely to achieve the same results as a third party company that works for several companies because it cannot obtain back loads or consolidate shipments to the same degree.[16] In some cases, companies turn to third party logistics to achieve the desired customer service level without having to use the cash needed for other investment into new product research and development and market expansion.

Research has shown that cross docking in a third party environment is usually initiated in two ways: through a cost-saving analysis initiated by the third party provider or at the request of third party customers. Of the two, a customer request is the primary initiator of cross docking programs.

- *Third Party Providers as CDO.* There are many examples of third party cross docking. Typically, multiple suppliers receive orders with allocation either at the facility level or by specific store. These suppliers ship their product to a third party provider's facility where pallets are broken down and store-specific pallets are built for customer's stores or based on geographic destination.

**REAL WORLD EXAMPLE**

**Exel Logistics Cross Docks In Mexico**

Exel Logistics—Consumer Sector operates many cross dock centers across the country and around the world. Most of the time the move to cross docking is initiated by the customer in an attempt to improve cycle time, reduce inventories, improve cost effectiveness, and reduce damage. One other reason is to take advantage of geography. Its facility in Mexico is centrally located, thus multiple suppliers send product to the facility where pool shipment occurs. Product becomes part of a dynamic pick line as soon as it arrives. Orders for specific customers are consolidated and loaded onto outbound trailers. By the end of the day, the dock is pretty much cleared of product in preparation for next day's shipment. Products stay in the center from 4 to 5 hours. With this cross dock program, they were able to reduce cycle time for its customer.

In the automotive industry, third party providers provide JIT support to manufacturing and assembly plants by consolidating component parts and raw materials from many vendors and cross docking them through their facilities.

- *Third Party Providers as Supplier*. In other cases, third party providers assume the role of a supplier by offering cross docking support to a CDO. Third party providers create combined loads made up of product from multiple suppliers/manufacturers so customers receive products from one source. These truckload quantities of multi-supplier unit loads are then shipped from the third party's facility to the customer's facility where they are cross docked to the customer's stores. By outsourcing, suppliers pool their resources, without incurring any major capital investment, while supporting the cross docking needs of their customers.

**Third Party Support for Spartan Stores**

Spartan Stores in Grand Rapids, MI, has been one of the innovators of cross docking with its deli products. It depends on a third party company to build store-ready units that are shipped to a Spartan warehouse and married up with other products before being sent to the store. Their strategy is to increase sales and not spend money in inventory investment or warehouse space or technology by letting the third party provider absorb that cost. Their WMS knows what is coming in every door, knows where it is coming from, when it is supposed to be there and where it is going, and is able to flash that information to the person accountable for filling the load. (Source: "Speeding Through the Warehouse," *WERCSheet*, July/August 1999, p. 4.)

REAL WORLD EXAMPLE

Facility, equipment, and information system requirements at the facility level remain the same whether the CDO is third party or not.

## 3.2 Step 1: Generate cross docking system designs

Generating appropriate cross docking system designs is not always difficult. If suppliers agree to consolidate products for the CDO (pre-allocated supplier consolidation), then the cross dock design is limited to a few system alternatives using mobile equipment to move pallets from receiving to shipping. If sortation is required, then the cross dock team can assemble multiple alternatives with variations in the flow and layout, equipment, degree of automation, programming routines by the WMS, and other variables. For each alternative, throughput requirements and recommendations as outlined in the previous chapter must be taken into consideration and matched with the different cross docking concepts presented earlier.

There are very few "pure" cross docking operations. In reality, some blend of cross docking and traditional warehousing occurs. In the initial stages of a cross docking program, only a few products tend to be cross docked and some products and suppliers do not have the potential for cross docking.[17] But as the CDO becomes more familiar with the concept and the trading partners realize its savings potential, cross docking should increase—and increase substantially. Therefore the cross dock team must design a facility that will be flexible enough to accommodate changes. Figure 44 illustrates combining cross docking with traditional warehousing functions in which cross docked merchandise merges with full case and broken case items picked from inventory and is sorted by store in an automated sortation system.

The following steps can guide the cross docking team in developing system designs (see Figure 45).

**FIGURE 44**

**Conceptual Layout with Cross Docking and Pick from Inventory Modules**

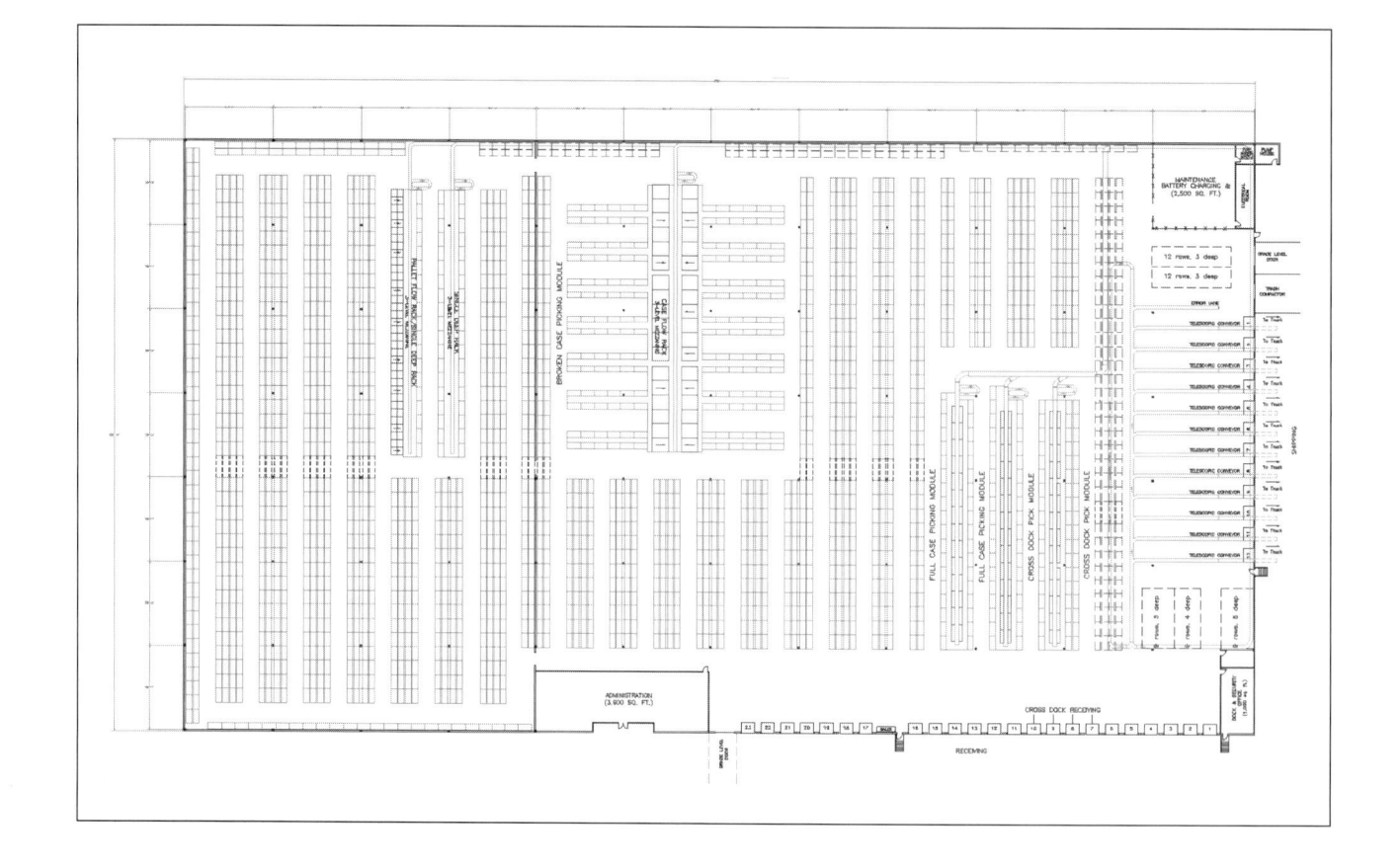

## Part 1: Assemble design parameters

The cross dock team needs to identify the following design parameters:

- Design year
- Defined planning horizons
- Projected throughput rate
- List of cross dock products and their suppliers
- Product characteristics
- Projected throughput rate per planning horizon
- Projected inventory eliminated with cross docked items
- Number of customers and customer profile
- Recommended customer delivery schedules, patterns, protocol and recommended supplier delivery schedules, patterns, protocol
- Sortation requirements (as listed in Chapter 2)

## Part 2: Develop alternative operational requirements

For each grouping of products (i.e., palletized loads versus carton quantities requiring sortation, pre-allocated versus post-allocated, labeled versus pre-labeled), develop the operations to carry out the cross docking of each group of SKUs. The chart shown on Figure 46 will help in developing alternatives for different scenarios. Product characteristics differ and may require a unique flow. Canada-based Oshawa Foods, a grocery wholesaler and retailer, made the move into large-scale cross docking by creating different operations for each product grouping. Instead of one distribution facility, the company went into partnership with Tibbett and Britten, a third party provider based in the UK, and operated a network of specialized centers that were integrated. In their fast-moving DC, a specific area, called a *mailbox*, was designated for each store where orders could be accumulated daily. Slow-moving items that were picked in bulk from a slow-moving warehouse, and produce that was picked in bulk from a perishables warehouse, are cross docked to this fast-moving DC and sorted into the store's mailbox. Orders for meats are pre-selected in the perishables warehouse and cross docked to the fast-moving DC, where they are kept in a cool storage staging area until all orders are ready to be shipped. They are then moved to marry with the rest of the orders for that store in preparation for truck loading. Thus for each product grouping, the company developed a unique cross docking operation.[18]

In generating alternatives, also consider existing equipment. Having a sortation system already in place that matches the required

**FIGURE 45**

**Phase 2, Step 1: Generate System Designs**

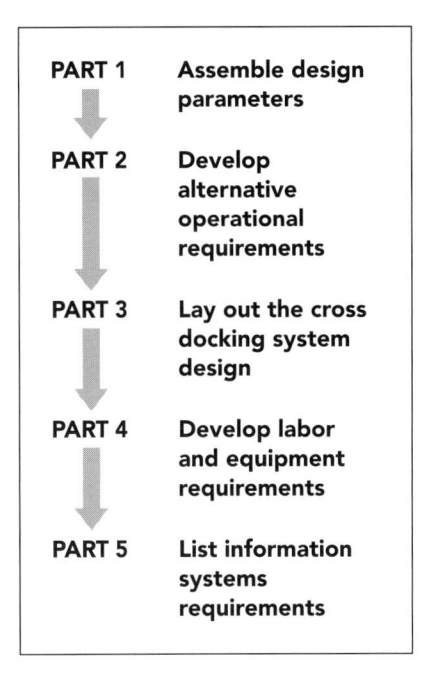

| PART 1 | Assemble design parameters |
| --- | --- |
| PART 2 | Develop alternative operational requirements |
| PART 3 | Lay out the cross docking system design |
| PART 4 | Develop labor and equipment requirements |
| PART 5 | List information systems requirements |

parameters simplifies the application of cross docking. Perhaps all that is required would be a pick-to-belt area where cross docked cartons can be picked to a conveyor belt that merges with other items in inventory and are sorted by store. On the other hand, building a new facility provides a clean slate for an almost unlimited array of layout and equipment alternatives.

**FIGURE 46**

**Cross Docking Systems and Their Operating Scenarios**

|  | CROSS DOCK SYSTEM |  |  |
|---|---|---|---|
|  | **Pre-allocated supplier consolidation** | **Pre-allocated CDO consolidation** | **Post-allocated CDO Consolidation** |
| **Samples of Product** | • High-cube, high demand SKUs for biggest stores<br>• Floor-ready merchandise, or displays pre-assembled<br>• Pre-assembled orders from another supplier<br>• Promotional products<br>• Seasonal products | • High-cube, high demand SKUs for biggest stores<br>• Floor-ready merchandise, or displays pre-assembled<br>• Pre-assembled orders from another supplier<br>• Promotional products<br>• Seasonal products<br>• Mixed pallets or less than pallet store order quantity<br>• Any pre-labeled cartons | • Less than pallet store order quantity<br>• Staple products<br>• Raw components for continuous production |
| **External Reliance** | High | Medium | Low |
| **MANUAL** | | | |
| **Operation Options** | • Full pallet moves from receiving to shipping | • Sort-To-Pallet | • Sort-To-Pallet<br>• Pick-To-Pallet |
| **Equipment (Throughput per Unit)*** | • Electric Pallet Jack (20–30 pallets per hr)<br>• CB Lift Truck (10–15 pallets per hr)<br>• Narrow Aisle Reach Truck (10–12 pallets per hr) | • Electric Pallet Jack (30–80 cartons per hr) | • Electric Pallet Jack (30–80 cartons per hr) |
| **CDO Investment** | Low | Low | Medium |
| **MECHANIZED** | | | |
| **Operation Options** | • Automatic pallet sortation system | • Automatic carton sortation system | • Automatic carton sortation system |
| **Equipment (Throughput per Unit)*** | • Pallet Conveyor System (approx. 500 pallets per hr) | • Carton Conveyor System (4,000–8,000 cartons per hr) | • Carton Conveyor System (4,000–8,000 cartons per hr) |
| **CDO Investment** | High | High | High |

\* The throughput standards are broadly categorized and are intended as rough estimates of productivity expectations. Many factors can significantly affect results. Please use these figures cautiously. Throughput standards from "Rules of Thumb" brochure provided by Gross & Associates. Check www.GrossAssociates.com.

## Part 3: Lay out the cross docking system design

Laying out different alternatives graphically helps the cross dock team visualize and analyze flows, choose and place permanent equipment, allow adequate aisles for mobile equipment, anticipate any impediments to the operation, plan for congestion and bottlenecks, and examine adjacency relationships. It also provides an estimate of space required—a critical factor to those who want to incorporate cross docking in an existing facility. Even for a new facility, space budgets will need to be prepared before any site selection

CHAPTER 3   *Phase 2: Planning and Designing a Cross Docking System*   95

can begin. There should be adequate space for the following cross docking functions:

- Receiving
- Receiving staging (if needed)
- Labeling (if needed)
- Breakdown of pallet loads into cartons (if needed)
- Sorting cartons (if needed)
- Palletizing cartons (if needed)
- Shipping staging (if needed)
- Shipping

Because all of these activities are concentrated around the receiving and shipping docks of a facility, the effective design of the dock space must be emphasized. Failure to plan carefully for the receiving and shipping needs of a cross dock facility may result in a congested dock area, costly future renovations, and equipment problems. In addition, it may cost the CDO demurrage for delaying carriers and impeding efficient receiving and shipping operations, thus compromising the cross docking system. Appendix A Provides information on designing docks.

A new facility offers a clean slate for the design team. Companies that anticipate cross docking have specified 60- to 70-foot dock depths in their new warehouses.[19] These deeper docks indicate the industry's acknowledgment of cross docking's savings potential. Within this dock, the amount of shipping staging will be impacted greatly by adopting a door-per-store concept. Door-per-store implies that a facility dedicates a loading door to each store with a trailer always in place to receive material for that store. Because cross dockable products are received, processed, and immediately loaded into these waiting trailers, staging is not required.

However, most cross docking tends to start out in an existing facility and the design is limited and constrained by existing columns, walls, and other installed equipment. Removing racks and breaking down walls may be required.

## Part 4: Develop labor and equipment requirements

Labor and equipment requirements can be determined from the *throughput rate*, such as the number of cross docked pallets unloaded per day, and the *unit handling time* of either a pallet, carton, or piece. The unit handling time can be determined either from actual time studies, predetermined time standards or from benchmarks taken from other companies handling similar product lines. The WMS can also be a source of valuable labor standards information. A record of transaction times can be downloaded to determine the time to handle a pallet with a particular piece of equipment. Frequently, cross docking does not require hiring new personnel. Because there is less handling, some of the labor previously required for putaway, storage, and picking can be used for cross docking.

Assume that there are 200 pallets unloaded per day. Time studies and WMS records show that it takes an average of 3.2 minutes to move a pallet with a counterbalanced lift truck from the inbound trailer to the outbound trailer. It will take 640 minutes to complete the work. Based on 7 working hours per person per day, it will require 1.52 persons ([640 min. ÷ 60 min./hr.]/7 hr.) to complete the job. To simplify the model, the number is rounded to 2 persons. Specialized computer software programs are available that include a data base of warehouse time standards for calculating productivity and labor requirements.[20]

### Part 5: List information systems requirements

For each alternative, determine the necessary information systems changes and upgrades. The previous section described the information system requirements for each type of cross docking system. Will minor programming changes be required? Or will a completely new WMS module for cross docking be needed? Will automatic sortation systems be used? If so, this will require scanners and PLCs or PCs able to communicate with the WMS. Will radio frequency devices be required? Will an automatic labeler be needed? By listing these requirements for each alternative, the team will be able to compare cross dock system costs.

## 3.3 Step 2: Perform an economic analysis on the alternatives

Analysis begins with determining estimated capital and annual operating costs for each cross docking concept. After all of the costs are assembled, simple but reliable methods for economically evaluating alternatives are used to determine the final alternative. The selection of the final alternative, however, will not depend solely on this economic analysis. It will also rely on a qualitative, supplementary analysis.

In conducting an economic comparison of cross docking alternatives, the following steps are suggested (Figure 47).

### Part 1: Assemble mutually exclusive alternatives

In the context of material handling, the alternatives might involve one or more conveyor layout designs or one or more mobile equipment alternatives. It is important to compare "apples to apples." All

**FIGURE 47**

Phase 2, Step 2: Perform an Economic Analysis on the Alternatives

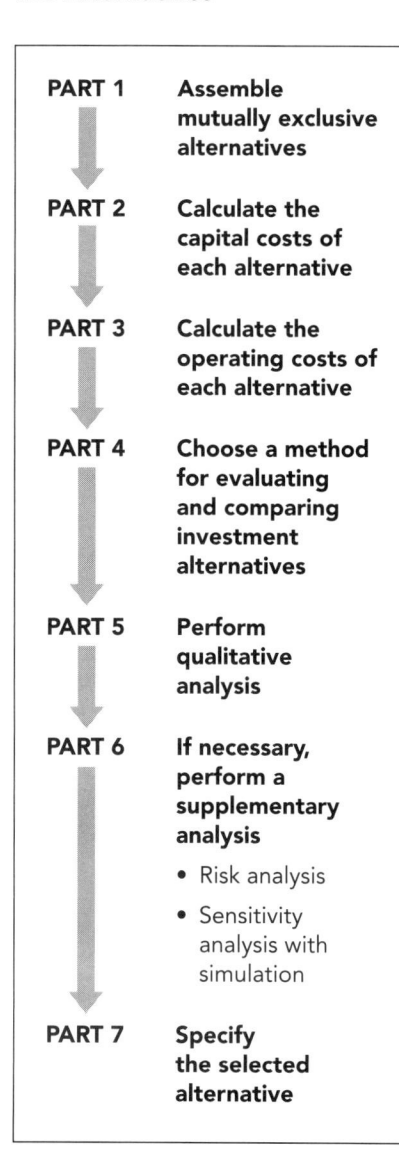

| PART 1 | Assemble mutually exclusive alternatives |
| PART 2 | Calculate the capital costs of each alternative |
| PART 3 | Calculate the operating costs of each alternative |
| PART 4 | Choose a method for evaluating and comparing investment alternatives |
| PART 5 | Perform qualitative analysis |
| PART 6 | If necessary, perform a supplementary analysis<br>• Risk analysis<br>• Sensitivity analysis with simulation |
| PART 7 | Specify the selected alternative |

CHAPTER 3   Phase 2: Planning and Designing a Cross Docking System

of the alternatives under evaluation should be within the same planning horizon and should be cross docking the same percentage of SKUs and the same volumes.

## Part 2: Calculate the capital costs of each alternative

The team needs to approximate the capital investment for each viable alternative. These are the initial costs to purchase, install, and implement equipment, facility, and information systems for system start-up and includes:

- Material handling equipment (e.g., lift trucks, conveyors, pallet jacks, etc.)
- Staging racks
- Facility upgrades or re-design required for each alternative (e.g., removal of racks, removal of walls, etc.)
- Information systems
  - Hardware (e.g., radio frequency equipment, printers, scanners, conveyor controls, communications equipment, etc.)
  - Software (e.g., development of software package, applications, operating systems, utility software, communications software)
- Any other costs related to the start-up of a cross docking system alternative

Capital costs that are common to all of the alternatives (such as training costs and the costs of hiring external consultants) are not included in this initial comparison of multiple alternatives. (All other costs related to the start-up of the selected cross docking system will be discussed in the next chapter.)

Figure 48 details approximate costs of common cross dock equipment as of 1999 for preliminary economic analysis. *These estimates may or may not be accurate for a particular application.* Regional cost-of-living differences and special requirements demanded by an application can

**FIGURE 48**

**Estimated Costs of Common Cross Docking Equipment**

Source: *Rules of Thumb*, Gross & Associates (www.GrossAssociates.com or 732-636-2666).

| EQUIPMENT CATEGORY | UNIT PRICE Lower Range | UNIT PRICE Upper Range |
|---|---|---|
| **Electric Trucks (with Battery & Charger)** | | |
| Walkie Pallet Jack | $5,500 | $7,500 |
| Rider Pallet Jack | $7,500 | $9,500 |
| Counterbalanced Truck | $21,000 | $29,000 |
| Narrow Aisle Reach Truck | $25,000 | $36,000 |
| Double Deep Reach Truck | $28,000 | $38,000 |
| **Miscellaneous Handling Equipment:** | | |
| Manual Pallet Truck | $400 | $900 |
| Scissor Lifts/Pallet Positioners | $1,200 | $3,600 |
| **Pallet Rack:** | | |

*(Rack priced per pallet positions, including floor positions. Assume 4 to 5 pallets high with standard 48" x 40" pallets with maximum load of 2,500 lbs per pallet. Assume structural or roll formed steel).*

| | | |
|---|---|---|
| Standard Selective Rack | $30 | $50 |
| Double Deep Rack | $40 | $55 |
| Drive-In Rack (4+ deep) | $50 | $65 |
| Drive-Thru Rack (4+ deep) | $55 | $75 |
| Gravity Flow Rack | $250 | $350 |
| **Conveyor:** | | |
| Live Roller (Per Foot) | $200 | $275 |
| Zero Pressure Accumulation (Per Foot) | $225 | $325 |
| Belt (Per Foot) | $180 | $250 |
| Gravity Roller/Skate (Per Foot) | $25 | $50 |
| Flexible Skate Wheel (Per Foot Extended) | $50 | $75 |

*The preceding budgetary estimates are installed prices. For substantial engineered conveyance projects, the following percentage breakdowns may apply (estimates only):*

| | |
|---|---|
| • Controls Engineering | 6% |
| • Mechanical Engineering | 7% |
| • Project Management | 3% |
| • Electrical Installation | 8% |
| • Control Hardware | 9% |
| • Mechanical Systems | 14% |
| • Mechanical Installation | 13% |
| • Mechanical Hardware | 40% |
| **TOTAL** | **100%** |

**FIGURE 49**

**Equipment Costs of Operation and Ownership Worksheet**

Source: Napolitano, Maida, *The Time, Space & Cost Guide To Better Warehouse Design*, Distribution Center Management, 1994.

result in significant differences in these estimates. In addition, union and non-union labor costs can significantly affect the cost of equipment installation.

## Part 3: Calculate the operating costs of each alternative

Identify estimated annual operating costs for each concept. For this analysis, include only the operating costs incurred specifically for the cross docking operation such as:

| CATEGORY | Example<br>VNA Turret Truck | Type No. 1 | Type No. 2 | Type No. 3 |
|---|---|---|---|---|
| **1. Capital costs** | | | | |
| a. Unit | $ 88,000 | | | |
| b. Battery | $ 8,200 | | | |
| c. Charger | $ 4,000 | | | |
| Total: | $ 100,200 | | | |
| **2. Years depreciation** | | | | |
| a. Unit | 25 | | | |
| b. Battery | 10 | | | |
| c. Charger | 20 | | | |
| **3. Expected yearly usage (hours)** | 2,000 | | | |
| **4. Depreciation costs** | | | | |
| a. Unit (line1a/ line 2a) | $ 3,520 | | | |
| b. Battery (line 1b/line 2b) | $ 820 | | | |
| c. Charger (line 1c/line 2c) | $ 200 | | | |
| Total: | $ 4,540 | | | |
| **5. Interest rate** | 7% | | | |
| **6. Interest cost per year (line 1 X line 5)** | $ 7,014 | | | |
| **7. TOTAL OWNERSHIP COSTS (line 4 + line 6)** | $ 11,554 | | | |
| **8. Gas, oil, or electricity yearly consumption** | | | | |
| Gas consumption (1 gallon per hour) | n/a | | | |
| Oil consumption (0.04 quart per hour) | n/a | | | |
| Electricity (per kilowatt hour) | 4,180 | | | |
| **9. Gas, oil, electricity cost per unit** | | | | |
| Gasoline (per gallon) | $ 0.99 | | | |
| Oil consumption (per quart) | $ 1.20 | | | |
| Electricity (per kilowatt hour) | $ 0.12 | | | |
| **10. Gas,oil, electricity cost per year (line 8 X line 9)** | | | | |
| Gasoline | n/a | | | |
| Oil | n/a | | | |
| Electricity | $ 502 | | | |
| **11. Maintenance costs** | | | | |
| Electric - 1.5% of capital cost | $ 1,503 | | | |
| Gasoline and LPG - 5% of capital cost | n/a | | | |
| **12. TOTAL OPERATING COSTS (line 10 + line 11)** | $ 2,005 | | | |
| **13. TOTAL OWNERSHIP AND OPERATING COSTS** | | | | |
| a. Per year (line 7/ line 12) | $ 13,559 | | | |
| b. Per hour (line 13a / line 3) | $ 6.78 | | | |

CHAPTER 3   *Phase 2: Planning and Designing a Cross Docking System*

- Labor costs related to each alternative
- Equipment costs related to each alternative
- Any unique maintenance, overhead, new systems support, utilities costs required by an alternative
- Operating space costs required by an alternative

Figure 49 provides a sample method for calculating annual equipment ownership and operating costs for specific material handling equipment. Consider only the space, labor, and equipment dedicated to the cross dock effort for this exercise. Operating costs that are common to all of the alternatives (such as inventory carrying costs, internal labor maintenance) are not included in this initial comparison. All other costs related to the operation and support of the selected cross docking system will be discussed in the following chapter.

## Part 4: Choose a method for evaluating and comparing investment alternatives

There are many measures of economic performance available, some of which are discussed briefly below.[21]

*Payback period method (PBP).* This method uses the simple concept that the net annual cash flow (NACF) derived from an investment should pay back the initial cost (IC) of the investment in a certain period of time. Assuming the net annual cash flow is equal from one year to the next, the payback period can be defined as follows:

$$PBP = \frac{IC}{NACF}$$

PBP = Payback period
IC = Initial cost of the investment
NACF = Net annual cash flow

*Example: If a semi-automated conveyor system for cross docking costs $750,000 (IC) installed and is expected to save approximately $210,000 (NACF), the payback period, PBP, is:*

$$PBP = \frac{\$750{,}000}{\$210{,}000}$$

$$= 3.57 \text{ years}$$

*In the analysis of alternatives, the scenario with the lowest payback period is usually selected.* Although it is easy to compute, use, and explain, analysts should refrain from making the payback period method the method of choice because it is shortsighted in that it ignores the financial performance of an investment beyond the calculated payback period. Nevertheless, it does provide a rough measure of the liquidity of a project. It is recommended for use only for breaking ties after the following other two methods are used.[22]

*Equivalent uniform annual cost method (EUAC).* This method converts all present and future costs of each alternative to their equivalent annual costs using an appropriate interest rate. The interest rate is the minimum attractive rate of return (MARR) that a company uses to evaluate investment opportunities. This rate is usually a policy matter agreed upon by the senior management of an organization. Factors influencing this rate include availability of funds for investment, competing investment opportunities, differences in the risks of competing investment opportunities, difference in the time required for recovery of the investment, and current interest rates of primary lenders.[23] Using this interest rate, annualized cash costs are determined for each alternative, and *the investment with the lowest annual costs would be the preferred alternative.*

*Example: A semi-automated conveyor sortation system has an initial cost (IC) of $1,000,000, annual costs of $55,000, and a salvage value of $350,000 after 15 years.*[24] *Using a MARR of 10%, the annual costs, EUAC, would be:*

EUAC (1) = $1,000,000 (A/P, 10%, 15) + $55,000 − $350,000
         (A/F, 10%, 15)
= $1,000,000 (0.13147) + $55,000 − $350,000 (0.03147)
= $131,470 + $55,000 − $11,014
= $197,484

*When compared to a manual alternative of cross docking with initial costs of $105,300, annual costs of $193,000, and a salvage value of $10,000 after 20 years, the annual worth, AW, would be:*

EUAC (2) = $105,300 (A/P, 10%, 20) + $193,000 − $10,000 (A/F,
         10%, 20)[25]
= $105,300 (0.11746) + $193,000 − $10,000 (0.01746)
= $12,369 + $193,000 − $175
= $205,194

*The alternative with the lower EUAC is preferred. In this case, it would be the semi-automated alternative with an annualized cost of $197,484.*

*Rate of return (ROR),* also known as the return on investment (ROI), method. This rate of return is the interest rate that yields zero EUAC or present worth.[26] The value of the interest rate that drives the EAUC or PW to zero is determined through simple interpolation or extrapolation. Interest tables from engineering economy publications (not included in this publication) are required to use this method.

*Example: The return on invested capital if $750,000 is invested now in material handling equipment with annual savings of $210,000 over 10 years would be:*

$$\text{EUAC} = -\$750{,}000\,(A/P, i, 10) + \$210{,}000 = 0$$

$$(A/P, i, 10) = \frac{\$210{,}000}{\$750{,}000}$$

$$= 0.28$$

*Scanning interest tables at n = 10 for different values of i, (A/P, 30%, 10) = 0.32346 and (A/P, 20%, 10) = 0.23852. By interpolation, a rate of return of 24.9% is determined for the investment. If MARR decided upon by the company is less 24.9% then the investment is acceptable.*

*An ROR analysis should not be used to rank or compare alternatives unless an incremental analysis is used.* In fact, because of this requirement, using this method is more tedious than the EUAC method. But when used correctly, the same recommendation will result using the ROR method and the EAUC and present worth methods. The advantage with ROR is that no knowledge of an interest rate is required.[27] It is recommended that the EUAC method be used to select the alternative and the ROR method be used to determine the expected rate of return for the investment.[28]

## Part 5: Perform qualitative analysis

Selecting the best system does not end with the quantitative models of economic performance. Other factors have to be considered such as:

- Ease of implementation
- Ease of operation
- Ease of monitoring and managing the system
- Familiarity with certain equipment
- Known reliability of certain equipment
- Flexibility of the equipment and layout to accommodate shifts in volume and changes in the business

There may be a virtual deadlock as a result of the economic comparison. This additional analysis will help the team to select the final alternative with the least operational challenges.

## Part 6: If necessary, perform supplementary analysis

In some cases, a virtual deadlock may necessitate the need for supplementary analysis such as sensitivity analysis or risk assessment. Sensitivity analysis involves changing a range of possible values for certain key parameters and determining their effect on the design. Risk analysis involves changing certain parameters and determining the probabilities of their occurrence. Sensitivity analysis is easiest to do with a simulation model. In fact, in highly mechanized cross docking systems, the study team is advised to simulate the alternatives. With simulation, analysts can replicate the actual system on the computer using mathematical relationships and constructs. Alternative designs and various levels of throughput can be tested to anticipate the limits and weaknesses of a system. Because simulation is a complex tool, which can entail an additional $60,000 to more than $100,000 in planning and design budgets, it is often considered overkill for designs with low capital risks.

## Part 7: Specify the selected alternative

The final step in conducting an economic comparison is to specify the preferred alternative. This will be the design for which final costs and savings are calculated and presented to senior management and the cross dock trading partners for final justification.

The case study in Chapter 6 describes an actual operational design process and economic analysis for a hardware retailer.

## Notes

1 This classification is a variation from a report prepared by the Coca-Cola™ Retailing Research Council for the grocery industry by Mercer Management Consulting entitled *New Ways to Take Costs Out of the Retail Food Pipeline*, A Study Conducted for the Coca-Cola Retailing Research Council, 1992.
2 *Rules of Thumb*, brochure provided by Gross & Associates for warehousing and distribution equipment costs and throughput standards (p.8). For copies: check www.GrossAssociates.com.
3 For additional information, contact Retrotech, Inc., 610 Fishers Run; P.O. Box 586, Fishers, NY 14453-0586, (716) 924-6333. See also www.retrotech.com.
4 Ahead of the Rest with...Cross Docking. ACTIV Systems brochure provided by Retrotech, Inc., p. 2.
5 Schaffer, Burt, "Implementing a Successful Cross Docking Operation," *IIE Solutions*, October 1997, p. 35.
6 Wagar, Kenneth, "Cross Dock and Flow Through Logistics For the Food Industry," *Annual Conference Proceedings*, Council of Logistics Management, October 1995, p. 188.
7 Wagar, Kenneth, *loc. cit.*, p. 184.
8 Wagar, Kenneth, *loc. cit.*, p. 185.
9 If a bar coded shipping label is already attached to the individual cartons, then warehouse personnel simply scan the carton so that the WMS can direct the worker to the carton's appropriate staging lane.
10 Pictures provided by Rapistan Systems. For more information contact Rapistan Systems, 507 Plymouth Avenue NE, Grand Rapids, MI 49505-6098, (616) 913-6525 or www.rapistan.com.
11 "The flow-through concept: We don't store it—we ship it," *Modern Materials Handling*, June 1990, p. 66.
12 Blaser, Jim, *loc. cit.*, p. 3.
13 For wave picking, some companies keep a trailer of product at an inbound door and unload as needed. Demurrage charges can make this unfeasible. With your own fleet of trucks, this method of "trailer staging" ties up a dock door and can incur additional operating costs if there are insufficient doors.
14 Donald Patterson, "Pausing-In-Transit: A Distinctive Option in Distribution, "*Warehousing Forum*, The Ackerman Company, April 1999.

15 Cooke, James Aaron, "Cross-docking Software: Ready or not?" *Logistics Management*, October 1997, p. 58.
16 McLeod, Marcia. "Cutting Both Ways," *Supply Management*, London, April 1, 1999, p. 24.
17 See Chapter 2 to determine how to qualify products for cross docking.
18 Smith, Michael J., Based on an October 1995 transcript of a presentation on cross docking for the Council of Logistics Management.
19 Thayer, Warren, "Logistics: Life in the Fast Lane," *Frozen Food Age*, July 1999, p. 29.
20 Contact Consulting Services Company for software called WHAM which is a database of warehouse-specific time standards, (312) 944-1787.
21 We do not attempt to discuss the different methods in detail, but briefly highlight the more popular approaches. For more information, consult any engineering economy publication.
22 Tompkins, James, et al. *Facility Planning*, 2nd ed., New York: John Wiley & Sons, 1996, pp. 674-675.
23 Grant, Eugene, et al. *Principles of Engineering Economy*, 7th ed., New York: John Wiley & Sons, 1982, p. 163.
24 Salvage value is often represented by the worth of the equipment if sold at the end of the economic life of the investment.
25 For an explanation of the multiplier (A/F, 10%, 20) convention, refer to Grant, *Principles of Engineering Economy*.
26 Present worth (PW) method resembles the EUAC method except that the annual and future cash flows are converted to their present worth.
27 Lindeburg, Michael R. *Engineer in Training Review Manual*, California: Professional Publications, Inc., 1982, pp. 2-5.
28 Tompkins, James, et al. *Loc. cit.* p. 682.

# Phase 3: Identifying Costs and Savings of a Cross Docking System

**Objective of Chapter 4:** *To determine overall savings and the cost impact both system-wide and at the product level.*

**4.1** Step 1: Create a system-wide cost model and determine ROI

**4.2** Step 2: Create a product cost model and determine impact on product profitability

**4.3** Step 3: Calculate the supply chain costs and savings

# Phase 3: Identifying Costs and Savings of a Cross Docking System

Selecting the final alternative from Phase 2 paves the way for a detailed look at the actual costs and savings impact of a cross docking initiative. Typically the shift is from a traditional, inventory-based system to a selective cross docking system for a few (10 percent to 20 percent) SKUs. But some companies make the shift from a direct-store-delivery (DSD) program to cross docking. Then, there are those rare occasions where companies have made a complete turnabout by cross docking 80 percent to 90 percent of their SKUs. Notwithstanding the level of transition, some capital investment will usually be required to make the shift. The typical way this capital is released for implementation to start is for senior management to find the program *economically justified*.

An accurate justification of capital for cross docking, or any new strategy for that matter, involves creating cost models. Cost models assemble relevant cost components and their actual costs in a logical manner to provide insight on the relationship between benefits and costs. The model is often a collection of multiple spreadsheets whose values on one are often linked to formulas in another. The spreadsheets are maintained on the computer for efficiency. Identifying true costs is often considered one of the biggest challenges in the realm of logistics and physical distribution. Certain aspects are often difficult to quantify and assign costs. Many companies create their own cost standards and subsequent methods for justification. In this publication, two types of models are suggested with varying objectives.

The first type compares supply chain costs on a *system-wide basis*. This model uses a traditional macroscopic approach by calculating the total incremental cost difference for all activity centers affected by cross docking for each trading partner. Companies initially embarking on a cross docking program will find this model invaluable for determining and comparing the overall return on their investments. This method, however, does not identify the effect of cross docking on an individual SKU's profitability.

The second type of model is more focused in that it compares costs with and without cross docking on a *product-by-product basis*.

CHAPTER 4 *Phase 3: Identifying Costs and Savings of a Cross Docking System*  107

This model enables suppliers, CDOs, and customers for that particular SKU to understand how their function in the supply chain contributes or takes away from its profit margin. Because determining the return on investment *and* insight into a product's profitability are both significant measures of a strategy's effectiveness, typically both models are created.

## 4.2 Step 1: Create a system-wide cost model and determine ROI

A system-wide cost model is created to determine the annual costs and savings with and without cross docking. Because it is primarily used as a decision-making tool (and *not* for actual day-to-day record keeping and tracking of costs), it uses only the *incremental costs* of those cost elements affected by cross docking. The primary purpose of the system-wide cost model is to identify the *change* in total costs brought about by introducing cross docking to the current system.

By its very nature, cross docking cuts across many logistics functions and impacts their cost. Therefore, the system-wide cost model will also be cross-functional. Depending on the type of cross docking employed, there may be costs incurred by the suppliers, by the CDO, and by the customers. Each affected cost and savings must be documented for a true justification.

Here, the building of the system-wide cost model is described with a working example of a retail chain that is initiating a pre-allocated supplier consolidation cross dock in its existing facility. Suppliers have agreed to supply the retail chain with multi-SKU pallets for the retailer to cross dock through his facility. Because this is a simple full-pallet cross dock with no case-level sortation, the company does not anticipate any facility changes or the purchase of new equipment. They are, however, planning to add an additional dock door and install bar coding, radio frequency devices and EDI communication for a more efficient cross dock. Figure 50 outlines the process.

**FIGURE 50**

Phase 3, Step 1: Create the System-wide Cost Model and Determine ROI

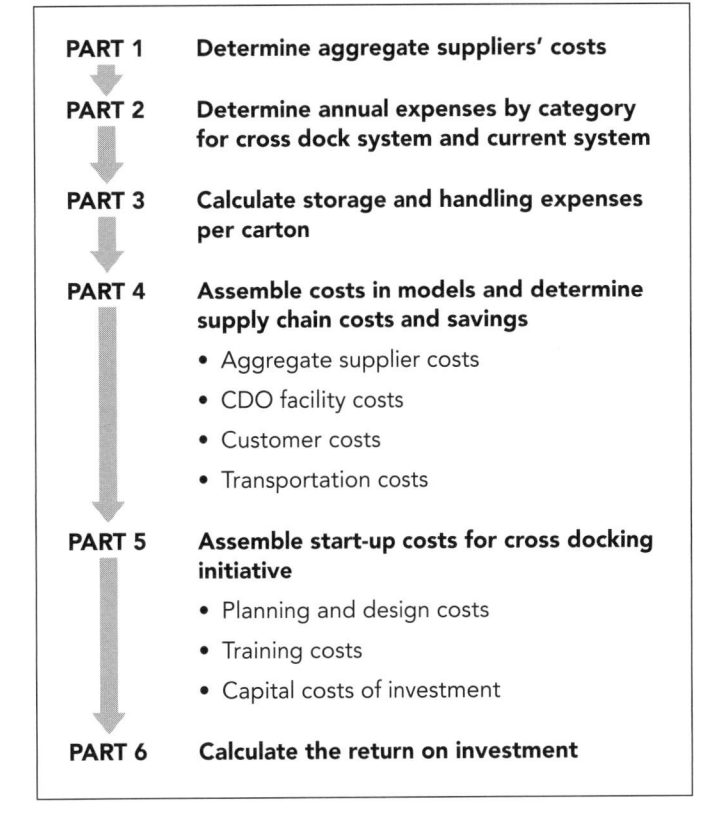

| | |
|---|---|
| PART 1 | Determine aggregate suppliers' costs |
| PART 2 | Determine annual expenses by category for cross dock system and current system |
| PART 3 | Calculate storage and handling expenses per carton |
| PART 4 | Assemble costs in models and determine supply chain costs and savings<br>• Aggregate supplier costs<br>• CDO facility costs<br>• Customer costs<br>• Transportation costs |
| PART 5 | Assemble start-up costs for cross docking initiative<br>• Planning and design costs<br>• Training costs<br>• Capital costs of investment |
| PART 6 | Calculate the return on investment |

### Part 1: Determine aggregate supplier costs

In pre-allocated supplier and CDO consolidation types of cross docking, the supplier's costs increase as a result of the program.

Where possible, the supplier should provide a fair and accurate accounting of these costs to the study team. It will provide the CDO with some perspective into the supplier's costs and its impact on the supply chain. At the supplier's facility, *handling costs* immediately come to mind. In addition, the supplier may incur additional *costs for storage* as an SKU may stay longer at a supplier's facility, awaiting assembly with other SKUs. There may also be additional *transportation costs* especially if the supplier is called upon to deliver fewer items more frequently.

Figure 51 lists a sample of costs incurred by different suppliers in our working example. For simplicity, only handling costs are considered. Supplier 1 incurs an additional cost of $0.41 per carton for mixing SKUs and additional labeling, but he keeps the acquisition price, or the price for the retail chain to purchase his product, unchanged at $50 per carton. Thus this supplier absorbs 100 percent of the costs. In contrast, Supplier 3 incurs an additional $0.60 in handling costs and passes half of the costs to the retail chain by increasing the acquisition cost from $30 to $30.30.

## Part 2: Determine annual expenses by category for the existing system and the cross dock system

While suppliers assemble their costs, the CDO must also assemble his costs. Each company organizes its costs differently, but the cost

**FIGURE 51**

**Incremental Supplier Costs for Fictional Retail Chain**

| SUPPLIER NUMBER | Traditional | Cross Dock | Difference |
|---|---|---|---|
| **CD Supplier 1: Mixed SKUs and Labeling** | | | |
| a. Cartons sold | 50,000 | 50,000 | – |
| b. Handling cost/carton* | $ 0.27 | $ 0.68 | $ (0.41) |
| c. Net acquisition cost per carton | $ 50.00 | $ 50.00 | $ – |
| Total Costs for Supplier 1 | $ 13,500 | $ 34,000 | $ (20,500) |
| CDO Acquisition Cost, Supplier 1 | $ 2,500,000 | $ 2,500,000 | $ – |
| **CD Supplier 2: Mixed Assortments and Labeling** | | | |
| a. Cartons sold | 35,000 | 35,000 | – |
| b. Handling cost/carton* | $ 0.31 | $ 0.41 | $ (0.10) |
| c. Net acquisition cost per carton | $ 45.00 | $ 45.00 | $ – |
| Total Costs for Supplier 2 | $ 10,850.00 | $ 14,350.00 | $ (3,500) |
| CDO Acquisition Cost, Supplier 2 | $ 1,575,000 | $ 1,575,000 | $ – |
| **CD Supplier 3: Store Ready Pallets and Labeling** | | | |
| a. Cartons sold | 60,000 | 60,000 | – |
| b. Handling cost/carton* | $ 0.27 | $ 0.87 | $ (0.60) |
| c. Net acquisition cost per carton | $ 30.00 | $ 30.30 | $ (0.30) |
| Total Costs for Supplier 3 | $ 16,200 | $ 52,200 | $ (36,000) |
| CDO Acquisition Cost, Supplier 3 | $ 1,800,000 | $ 1,818,000 | $ (18,000) |
| **Total Cartons Sold** | 145,000 | 145,000 | – |
| **Total CD Supplier Costs** | $ 40,550 | $ 100,550 | $ (60,000) |
| **Total CDO Acquisition Cost** | $ 5,875,000 | $ 5,893,000 | $ (18,000) |

* Includes expenses incurred by additional labor and equipment required to sort and label product before shipment to CDO. Values provided by Supplier.

CHAPTER 4 Phase 3: Identifying Costs and Savings of a Cross Docking System

components remain inherently the same. The model shown in Figure 52 is developed for calculating warehousing costs.[1] It isolates handling, storage, operating administration (OA) and general administration (GA) costs. Handling expenses include all the costs associated with the movement of product into and out of the warehouse. Storage expenses are incurred with the ownership (or rental), operation and maintenance of the facility regardless of whether or not any product in the warehouse ever moves. OA costs are incurred in supporting a warehouse operation and are generally tied to a specific warehouse. If the warehouse were closed, these

**FIGURE 52**

Annual Expenses by Category

|  | Traditional | Cross Dock | Difference |  |
|---|---|---|---|---|
| **DIRECT HANDLING EXPENSES** |  |  |  |  |
| Equipment: |  |  |  |  |
|     Equipment depreciation and interest | $ 355,400 | $ 269,960 | $ 85,440 | less equipment to operate & maintain |
|     Utilities | 13,000 | 9,900 | 3,100 |  |
|     Maintenance | 70,100 | 53,300 | 16,800 |  |
|     Supplies and parts: handling equipment | 45,400 | 34,000 | 11,400 |  |
| Labor: |  |  |  |  |
|     Hourly wages | $ 1,301,000 | $ 988,500 | $ 312,500 |  |
|     Hourly benefits | 180,500 | 136,600 | 43,900 |  |
|     Purchased labor | 52,000 | – | 52,000 |  |
|     Hourly payroll taxes | 289,500 | 220,000 | 69,500 | less labor required |
|     Paid vacations/holidays: hourly | 24,000 | 18,000 | 6,000 |  |
|     Other handling expenses | 41,600 | 32,740 | 8,860 |  |
| **Total Direct Handling Expenses** | **$2,372,500** | **$1,763,000** | **$ 609,500** |  |
| **DIRECT STORAGE EXPENSES** |  |  |  |  |
|     Facility depreciation and interest | $ 665,100 |  |  |  |
|     Real estate taxes | 27,000 |  |  | no change in storage; no upgrades anticipated |
|     Utilities | 186,200 |  |  |  |
|     Insurance: facility | 34,400 |  |  |  |
|     Exterior maintenance/grounds | 28,600 |  |  |  |
| **Total Direct Storage Expenses** | **$ 941,300** |  |  |  |
| **OPERATING & GENERAL ADMINISTRATIVE EXPENSES** |  |  |  |  |
| Operating: |  |  |  |  |
|     Supervisory salaries and fringes | $ 210,000 |  |  |  |
|     Clerical supervision: salaries & fringes | 80,000 |  |  |  |
|     Off-site travel: whse/sup. management | 30,000 |  |  |  |
|     Training: management & supervisory | 45,000 |  |  | no change in O&GA |
|     Information processing/computer | 60,000 |  |  |  |
| General: |  |  |  |  |
|     Exec. officer salaries & fringes | $ 130,000 |  |  |  |
|     Exec. officer travel | 40,000 |  |  |  |
|     Selling expense | 65,000 |  |  |  |
| **Total Operating & General Expenses** | **$ 660,000** |  |  |  |
| **GRAND TOTAL** | **$3,973,800** | **$3,364,300** | **$ 609,500** |  |

expenses would disappear. On the other hand, GA costs are not directly related to the operation of a specific warehouse. Instead they support the overall mission of the company.

The projected costs with cross docking should also be determined. In the previous phase (Chapter 3), labor and equipment operating cost requirements with the new strategy were already determined. The savings in costs should be incorporated with other existing costs. Keep in mind that some products will still be handled the traditional way with only a few SKUs making the shift to cross docking. Handling, storage, OA and GA expenses should reflect this change where applicable. In our example, fewer workers, lift trucks and electric pallet jacks will be needed in the new operation. There is no expected change in storage, OA, and GA expenses.

## Part 3: Calculate storage and handling expenses per carton

Figure 53 illustrates a method for calculating storage and handling expenses per carton. To better identify the true costs of each activity, some companies use activity-based costing in developing handling costs for each activity in the supply chain. Although it provides a more accurate picture of costs, such detail is not called for in this

**FIGURE 53**

**Calculating Storage and Handling Expenses Per Carton**

|  | Traditional | Cross Dock |
|---|---|---|
| **VARIABLES FOR CALCULATION FROM FIGURE 52** | | |
| a. Direct Handling Expenses | $ 2,372,500 | $ 1,763,000 |
| b. Direct Storage Expense | $ 941,300 | $ 941,300 |
| c. Operating & General Admin. Expenses | $ 660,000 | $ 660,000 |
| d. Grand Total | $ 3,973,800 | $ 3,364,300 |
| **SPREADSHEET A: CALCULATING STORAGE EXPENSES PER CARTON IN INVENTORY** | | |
| 1. Direct Storage Expenses (B) | $ 941,300 | |
| 2. 50 percent of OA & GA total (50% × C)† | $ 330,000 | |
| 3. Total Storage Expense (line 1 + line 2) | $ 1,271,300 | |
| 4. Total Gross Square Feet (size of facility) | 120,000 | |
| 5. Storage Expenses per Gross Square Foot (line 3 ÷ line 4) | $ 10.59 | |
| 6. Square Feet Per Pallet* | 7.1 | |
| 7. Cartons Per Pallet | 24 | |
| 8. Square Foot Per Carton (line 6 ÷ line 7) | 0.3 | |
| 9. Storage Expenses Per Carton (line 5 × line 8) | $ 3.13 | |
| **SPREADSHEET B: CALCULATING HANDLING EXPENSES PER CARTON IN INVENTORY** | | |
| 1. Direct Handling Expenses (A) | $ 2,372,500 | $ 1,763,000 |
| 2. 50 percent of OA and GA total (50% × C)† | $ 330,000 | $ 330,000 |
| 3. Total Handling Expenses (line 1 + line 2) | $ 2,702,500 | $ 2,093,000 |
| 4. Annual Throughput in Cartons | 1,200,000 | 1,200,000 |
| 5. Handling Expenses Per Carton (line 3 ÷ line 4) | $ 2.25 | $ 1.74 |

\* Based on inventory bulk space calculations for products to be cross docked.

† As determined by the company.

particular model, because our goal is not to isolate specific savings but to determine overall system-wide savings.

## Part 4: Assemble costs in the model and determine supply chain costs and savings

The cost elements for determining *total costs* should include (Figure 54):

1. *Aggregate Supplier Costs* includes all of the incremental costs incurred by the supplier for providing value-added product to enable cross docking at the next level of the chain.

2. *CDO Facility Costs* is a collection of costs incurred by the CDO. (Formulas are presented as guides.) Each company may calculate these costs differently.

   a. Total Handling Costs represents the annual costs associated with handling products moving through the facility.

   [Total Cartons Sold × Handling Expense/Carton]

   b. Total Storage Space Costs represents the annual costs associated with storing inventory in the facility.

   [Total Cartons in Inventory × Storage Expense/Carton]

   c. Capital Cost of Inventory Investment is the opportunity costs of carrying inventory when the capital could be invested elsewhere.

   [Total Acquisition Costs (from all suppliers) × Daily Interest Rate × Days of Inventory Paid (including days while product is in transit)]

   d. Inventory Service Costs consist of taxes and insurance paid as a result of holding inventory. The percentage used is usually calculated from last year's ratio of taxes plus insurance to the value of inventory. With cross docking this cost is reduced by reducing the quantity of merchandise held in inventory.

**FIGURE 54**

**Costs for System-wide Model**

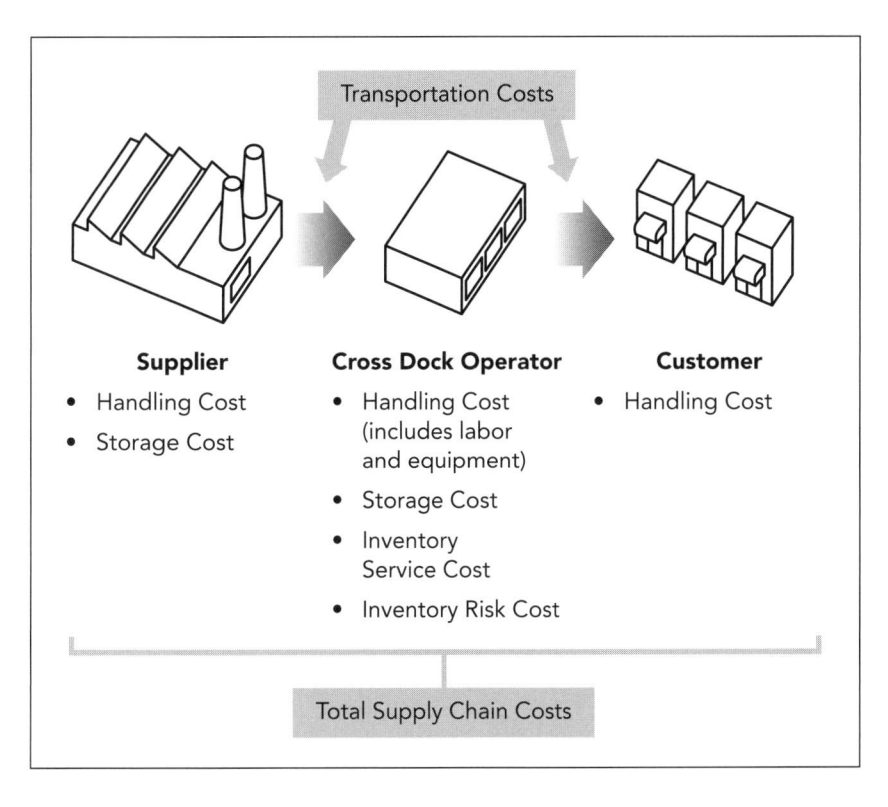

Transportation Costs

**Supplier**
- Handling Cost
- Storage Cost

**Cross Dock Operator**
- Handling Cost (includes labor and equipment)
- Storage Cost
- Inventory Service Cost
- Inventory Risk Cost

**Customer**
- Handling Cost

Total Supply Chain Costs

[Total Acquisition Costs × percent of Value for Insurance & Taxes × Average Year in Inventory]

  e. Inventory Risk Costs can vary from company to company but typically include charges for obsolescence, pilferage, and damage. It is usually expressed as last year's ratio of the quantity of obsolete, pilfered, and damaged product to the value of inventory. Because the amount of time that product is held as inventory is reduced or eliminated with cross docking, the cost for obsolescence is also reduced. Dramatic reductions in handling with some cross dock programs also diminish the instances of product damage and pilferage.

  [Total Acquisition Costs × percent of Value for Obsolescence, Damage, & Shrinkage × Average Year in Inventory]

3. *Customer Costs* are expenses impacted by cross docking at the store or plant or end-destination to which the cross docked product is delivered. In retail, cross docking custom pallets can result in a significant reduction of handling costs at the store. Pallets are delivered for immediate display on the store floor; no restocking is required to put the product on the shelf. In other cases, the cross docked, multi-SKU pallet is assembled according to the store layout, which also improves restocking productivity and reduces handling costs.

  [Customer Handling Expense Per Carton × Cartons Sold]

4. *Transportation Costs.* Inbound and outbound transportation costs differ for each company and often depend on the type of carrier used—whether is it private, common, or contract. Cross docking may or may not affect transportation costs. The same number of cartons will be moving through the warehouse. The same outbound loads will be created. It is in the timing that cross docking has an impact. Instead of multiple trucks arriving from one supplier once a month, the trucks may arrive once a week, four times a month depending on a customer's delivery schedule. In some instances, however, trucks may be required to do more frequent deliveries in fewer quantities. Therefore, one can expect a subsequent increase in these costs.

  [Cartons Sold × Average Transportation Cost Per Carton]

  *Note:* Inbound transportation costs from suppliers may have already been included with the supplier costs calculated in (1a). The study team should be careful not to double up on these costs. Verify with the suppliers to determine that their transportation costs have been included.

## FIGURE 55

**Sample Model for Determining Cost Savings Across the Supply Chain**

| COST ELEMENTS | Traditional | Cross Dock | Difference |
|---|---:|---:|---:|
| **Parameters Used in Calculation** | | | |
| A  Cartons Sold | 145,000 | 145,000 | – |
| B  Average Days in whse inventory | 20 | 2 | 18 |
| C  Average Year In inventory (line B ÷ 365) | 0.055 | 0.005 | 0.049 |
| D  Cartons in Inventory (C × A) | 7,945 | 795 | 7,151 |
| E  Transit Time in Days | 1 | 1 | – |
| F  Days of Inventory Paid (B + E) | 21 | 3 | 18 |
| G  Handling expense/carton (Form B) | $ 2.25 | $ 1.74 | $ 0.51 |
| H  Storage expense/carton (Form A) | $ 3.13 | $ 3.13 | $ – |
| I  Restocking expense/carton at store | $ 1.24 | $ 0.83 | $ 0.41 |
| **AGGREGATE SUPPLIER COSTS** | $ 40,550 | $ 100,550 | $ (60,000) |
| **RETAIL CHAIN COSTS** | | | |
| **Total Handling Costs (A × G)** | $ 326,552 | $ 252,904 | $ 73,648 |
| **Total Storage Space Costs (D × H)** | $ 24,901 | $ 2,490 | $ 22,411 |
| **Capital Cost of Inventory Investment** | | | |
| Net Acquisition Costs | $ 5,875,000 | $ 5,893,000 | |
| × Daily Interest Rate | 0.0411% | 0.0411% | |
| × Days of Inventory Paid | 21 | 3 | |
| = Capital Cost on Inventory | $ 50,707 | $ 7,266 | $ 43,441 |
| **Inventory Service Costs** | | | |
| Net Acquisition Costs | $ 5,875,000 | $ 5,893,000 | |
| × % of Value for Insurance & Taxes | 3% | 3% | |
| × Average Year In Inventory | 0.055 | 0.005 | |
| = Inventory Service Costs | $ 9,658 | $ 969 | $ 8,689 |
| **Inventory Risk Costs** | | | |
| Net Acquisition Costs | $ 5,875,000 | $ 5,893,000 | |
| × % of Value for Obsoles., Damage, Shrink | 2.00% | 0.75% | |
| × Average Year In Inventory | 0.055 | 0.005 | |
| = Inventory Service Costs | $ 6,438 | $ 242 | $ 6,196 |
| **RETAIL CHAIN STORE COSTS** | | | |
| = Restocking costs (A × I) | $ 179,800 | $ 119,865 | $ 59,935 |
| **TOTAL RETAIL CHAIN COSTS** | | | |
| Net Acquisition Costs | $ 5,875,000 | $ 5,893,000 | $ (18,000) |
| + Total Handling Costs | $ 326,552 | $ 252,904 | $ 73,648 |
| + Total Storage Space Costs | $ 24,901 | $ 2,490 | $ 22,411 |
| + Capital Cost on Inventory | $ 50,707 | $ 7,266 | $ 43,441 |
| + Inventory Service Costs | $ 9,658 | $ 969 | $ 8,689 |
| + Inventory Risk Costs | $ 6,438 | $ 242 | $ 6,196 |
| + Restocking Costs | $ 179,800 | $ 119,865 | $ 59,935 |
| = Total Operating Costs to CDO | $ 6,473,056 | $ 6,276,737 | $ 196,320 |
| **TOTAL SUPPLY CHAIN COSTS** | | | |
| Supplier Costs (1) | $ 40,550 | $ 100,550 | $ (60,000) |
| + Total Retail Chain Costs (4) | $ 6,473,056 | $ 6,276,737 | $ 196,320 |
| = Total Supply Chain Costs or Savings | $ 6,513,606 | $ 6,377,287 | $ 136,320 |

Annotations:
- Additional Aggregate Costs incurred by suppliers (AGGREGATE SUPPLIER COSTS row)
- Savings by Retail Chain Stores (Restocking costs row)
- Total Savings by Retail Chain (Total Operating Costs to CDO row)
- Savings by Supply Chain from Cross Docking (Total Supply Chain Costs or Savings)

5. *Total CDO Costs* are the aggregate costs incurred by the CDO. If the CDO and the customer are one and the same company, the total costs should include costs at the store or customer level.

6. *Total Supply Chain Costs* include all the costs impacted by cross docking from the supplier to the customer.

Figure 55 (previous page) illustrates these costs for our sample retail chain and its stores. Savings from cross docking can be easily identified on each level of the chain. This particular model shows an increase in costs of $60,000 for the suppliers but $196,320 in total savings by the retail chain. In addition, the supply chain experiences a net savings of $136,320 annually with a shift to a cross docking strategy. The retailer may invest some or all of its logistics savings from cross docking to provide improved value to the end consumer and consequently improve the marketability of its supplier's products. The supplier can track these savings "reinvested" by the retailer and any subsequent increase in revenue. They can then use this revenue increase to determine the return on investment for assembling custom pallets. With an increase in revenue, perhaps a supplier's upper management will expand their cross docking support program and thus increase overall savings for the supply chain.[2]

## Part 5: Assemble start-up costs for cross docking initiative

To calculate the return on investment for the cross dock initiative, the capital at start-up will also need to be assembled. Included in these costs are the following:

- *Planning and Design Costs* are estimated expenses incurred by the study and implementation team to assess, design, justify, and implement the program.

- *Training Costs* are the cost incurred to develop training materials and conduct the actual training to acquaint the necessary personnel with the new concept.

- *Capital Costs of Investment* are the costs incurred with the purchase, delivery, and installation of material handling equipment and information systems. This also includes costs for any facility upgrades or design specifically for the cross docking program.

Figure 56 illustrates the start-up costs incurred by the retail chain. Logistics consultants were used to develop the cross dock operation and procedures, to act as liaisons between supplier and the chain, and to evaluate information systems equipment bids. In fact, they were also charged with the task of working with both the CDO and the supplier to develop true costs. Using an independent consultant to develop costs introduces a certain measure of fairness

CHAPTER 4  Phase 3: Identifying Costs and Savings of a Cross Docking System            115

| COST ELEMENTS | Labor Hours | Cost Per Hour | Total Costs |
|---|---|---|---|
| **Planning and Design Cost** | | | |
| Study & Implementation Team | | | |
|     Logistics Manager | 240 | $ 50 | $ 12,000 |
|     Operations Manager | 230 | $ 40 | $ 9,200 |
|     Procurement Manager | 250 | $ 40 | $ 10,000 |
|     Merchandising Manager | 150 | $ 40 | $ 6,000 |
|     Accounting Manager | 50 | $ 35 | $ 1,750 |
|     Information Systems Manager | 150 | $ 35 | $ 5,250 |
|     Engineers | 300 | $ 25 | $ 7,500 |
|     Others | 80 | $ 25 | $ 2,000 |
| External Consultants | | | |
|     Logistics Consultants | flat rate | N/A | $ 60,000 |
|     Information Systems Consultants | 300 | $ 200 | $ 60,000 |
| Training Costs | | | $ 30,000 |
| **Total Planning and Design Cost** | | | **$ 203,700** |

| Capital Costs of Investment | No. of Units | Cost Per Unit | Total Costs |
|---|---|---|---|
| Material Handling Equipment | | | $ – |
| Facility Redesign/Upgrade | | | $ 10,200 |
| Information Systems | | | $ 259,750 |
| **Total Capital Costs of Investment** | | | **$ 269,950** |

**FIGURE 56**

**Start-up Costs for Cross Docking Initiative**

into the negotiation process between supplier and CDO. Information systems consultants were used to develop the upgrade for the WMS. There was no need to purchase any additional material handling equipment. Adding a dock door and improving the docks were the only upgrades performed on the facility.

## Part 6: Calculate the return on investment

Research shows numerous methods for computing the return on investment or ROI. Chapter 3 provided brief examples and concerns with regards to the calculation of the ROI. Briefly, the ROI is the interest rate that would yield identical profits if all the money were invested at that rate. It should be greater than a company's MARR (Minimum Attractive Rate of Return: the interest rate that the company wants their money to earn). A common, but incorrect, method of calculating ROI involves simply dividing the annual receipts of benefits by the initial investment. This technique ignores depreciation and the time value of money and should be avoided. In our example, the ROI for the retail chain is calculated as follows:

    The annual savings from the cross dock is $196,320 (from Figure 55). Start-up costs total $473,650 ($203,700 + $269,950 from Figure 56). These present worth costs have to be converted to annualized costs if the EUAC is used to calculate the ROI. Salvage values are ignored for simplicity, but should be added to the equation if available. The company is looking at a planning horizon of 10 years to determine the return on its cross docking investment.

$$\text{EUAC} = -\$373{,}650\,(A/P, i, 10) + \$196{,}320 = 0$$

$$(A/P, i, 10) = \frac{\$196{,}320}{\$473{,}650}$$

$$= 0.41448$$

Scanning interest tables (not included in this publication) at n= 10 for different values of i, (A/P, 45%, 10) = 0.46123 and (A/P, 35%, 10) = 0.37519. By interpolation, a rate of return of 39.6% is determined for the investment. The chain's MARR is currently at 15%. By investing in cross docking, the company increases their ROI by 164%, and full support from senior management can be expected. With such a high ROI, management may even decide to fully implement the program without looking at individual product costs. However, a project may have a high ROI, and still not be justified for an individual product.

## 4.2 Step 2: Create a product cost model and determine impact on product profitability

The system-wide cost model provides overall savings with cross docking. The product cost model specifically looks at the impact of cross docking on an individual SKU. This is especially useful during negotiations with individual suppliers and customers, because it allows them to see the cost drivers that may or may not be under their control and that, in the end, affect each product's profitability. Two approaches are summarized. The first approach involves the use of direct product profitability, or DPP, an application of logistics cost analysis that has gained widespread acceptance in the retail industry. The second involves the use of Activity Based Costing, a relatively newer approach created to address growing dissatisfaction with traditional accounting methods.

### Direct Product Profitability

Before DPP, an item's profitability was measured by using gross profit measurement. This was simply the sales price minus the costs of goods sold. In the late 1980s, retailers realized that this type of profitability analysis did not really identify all the costs attached to the product as it moved through the supply chain. There are hidden costs that can sometimes be substantial, so that the net profit on a particular item is reduced or even eliminated. Direct product profitability was created to take into account *all of the direct costs in the retail chain*, providing a greater level of accuracy than gross profit

## FIGURE 57

**SKU Cost Model Utilizing Direct Product Profit Method**

Source: Martin Christopher, "Integrating Logistics Strategy in the Corporate Financial Plan," *The Logistics Handbook*, (New York: The Free Press) 1994, p. 256.

| | | |
|---|---|---|
| SKU Number/s: _____ | Length: _____ | Avg Inv: _____ |
| SKU Description/s: _____ | Width: _____ | |
| Supplier: _____ | Height: _____ | Inv. Turns: _____ |
| Special Requirements: _____ | Weight: _____ | |

| | CATEGORIES | Traditional Warehousing | Cross Docking |
|---|---|---|---|
| | Sales | | |
| − | Cost of Goods Sold | | |
| = | Gross Margin | | |
| + | Allowances and Discounts | | |
| = | Adjusted Gross Margin | | |
| − | Total Warehouse Costs | | |
| | labor (labor model – case, cube, weight) | | |
| | occupancy (space and cube) | | |
| | inventory (average inventory) | | N/A |
| − | Total Transportation Costs (cube) | | |
| − | Total Retail Costs | | |
| | stocking labor | | |
| | front-end labor | | |
| | occupancy (cube) | | |
| | inventory (backroom) | | |
| = | **Direct Product Profit** | | |

All cost figures should be rationally allocated or assigned to an individual product.

measurement. DPP is the contribution to profit of an item that is calculated by:

- Adjusting the gross margin for each item to reflect deals, allowances, net forward buy income, prompt payment discounts, etc.

- Identifying and measuring the costs that can be directly attributed to individual products like labor, space, inventory, and transport.[3]

The model in Figure 57 describes the steps in moving from a crude gross margin measure to a more precise DPP. This model can also be used to compare the DPP of a product that is warehoused versus the same product if it were cross docked. It is crucial to look at DPP at the item level—especially for a cross dock justification analysis. Product characteristics, such as the cube, weight, space occupied by inventory, turns, and handling requirements, can vary considerably by item and affect the associated costs. All of the parties involved in the cross docking effort can use this cost model to determine specifically where they can achieve savings or incur costs.

## Activity-Based Costing

In the previous method, costs were assigned to specific cost centers. In ABC, costs are traced to specific activities and the cost drivers

**FIGURE 58**

Activity-Based Costing Concepts

consumed by each activity. ABC manages costs and operations data so that **resources** (utilities, labor, equipment, supplies, supervision) are related to **activities** (receiving, putaway, storage, order picking, shipping) and activities are related to **cost objects** (customers, products, product divisions) using multiple **cost drivers** (handling hours, number of pallets, number of lines picked). Figure 58 illustrates a conceptual ABC model and defines key terminology.

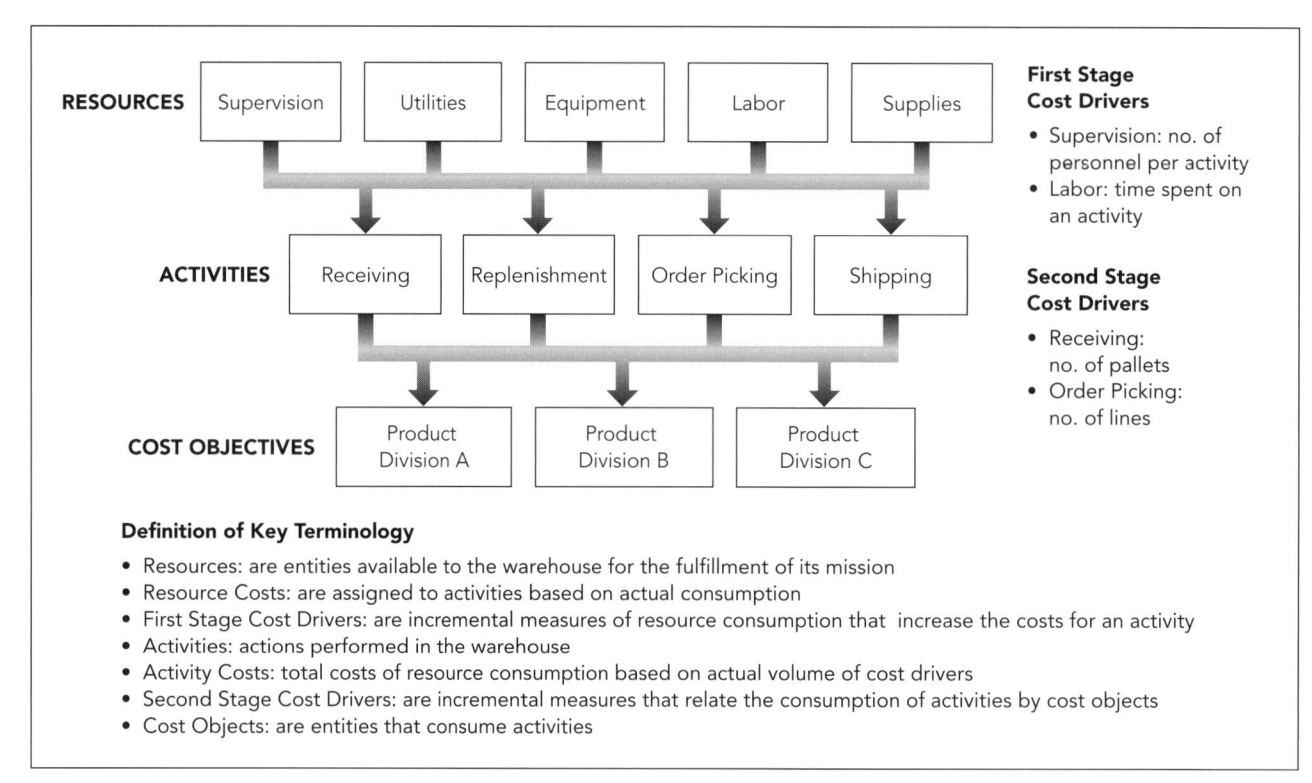

**RESOURCES:** Supervision, Utilities, Equipment, Labor, Supplies

**ACTIVITIES:** Receiving, Replenishment, Order Picking, Shipping

**COST OBJECTIVES:** Product Division A, Product Division B, Product Division C

**First Stage Cost Drivers**
- Supervision: no. of personnel per activity
- Labor: time spent on an activity

**Second Stage Cost Drivers**
- Receiving: no. of pallets
- Order Picking: no. of lines

**Definition of Key Terminology**
- Resources: are entities available to the warehouse for the fulfillment of its mission
- Resource Costs: are assigned to activities based on actual consumption
- First Stage Cost Drivers: are incremental measures of resource consumption that increase the costs for an activity
- Activities: actions performed in the warehouse
- Activity Costs: total costs of resource consumption based on actual volume of cost drivers
- Second Stage Cost Drivers: are incremental measures that relate the consumption of activities by cost objects
- Cost Objects: are entities that consume activities

The first step toward converting traditional cost data for ABC analysis is identifying activities. *Activities are tasks performed on a regular basis such as receiving a product or picking an order.* These activities follow the flow of product as it moves through the facility and represent differences in costs absorbed by a class of product.

The second step is to identify the first-stage cost drivers so that resources can be assigned to activities. In Figure 59 the salaries were broken down according to the time individuals spent performing specific activities in the warehouse. Three employees spend 75 percent of their time unloading inbound trailers, so 75 percent of their salary (including taxes, benefits, etc.) is assigned to unloading trailers based on a first-stage cost driver of percentage of total labor time. For storage the driving factors that affect the costs is the cube of the item and its inventory turns. The number of pallet positions reserved for that product reflects these factors. Thus the number of pallet positions is the cost driver.

CHAPTER 4  *Phase 3: Identifying Costs and Savings of a Cross Docking System*   119

Other cost drivers can be identified for each activity. Inspection may be driven by a complexity factor based on engineered standards, storage by the number of pallet positions, staging by the number of staged pallets, and shipping by the number of orders. Jack Haedicke, director of activity-based costing for Coca-Cola™ Co., gives the following example.[4] Inspecting paper products is simple. There are a few items to a pallet, all the items are the same, and there is rarely any damage. Inspection might take 20 seconds per pallet. A pallet of health and beauty care items could literally take hours to receive. There are many SKUs on a pallet, each SKU could have dozens of items, and each item may have a high value. The cost driver should reflect this complexity. With the application of cross docking, only reliable vendors that provide consistent good quality items are included in the program, and costs attributed to the inspection activity should substantially decrease or be eliminated for some items.

The third step is to specify the cost object that consumes the activity. Since the cross dock team wants to evaluate costs on a product basis, the product, or a class of products possessing similar characteristics, becomes the cost object.

Next the second-stage cost driver should be identified. Second-stage cost drivers reflect how cost objects use activities. In Figure 59, one of the activities is unloading inbound trailers. The cost for this activity for a class of products was driven by the number of pallets unloaded, so the number of pallets unloaded was selected as the second-stage cost driver.

**FIGURE 59**

**ABC Model Example**

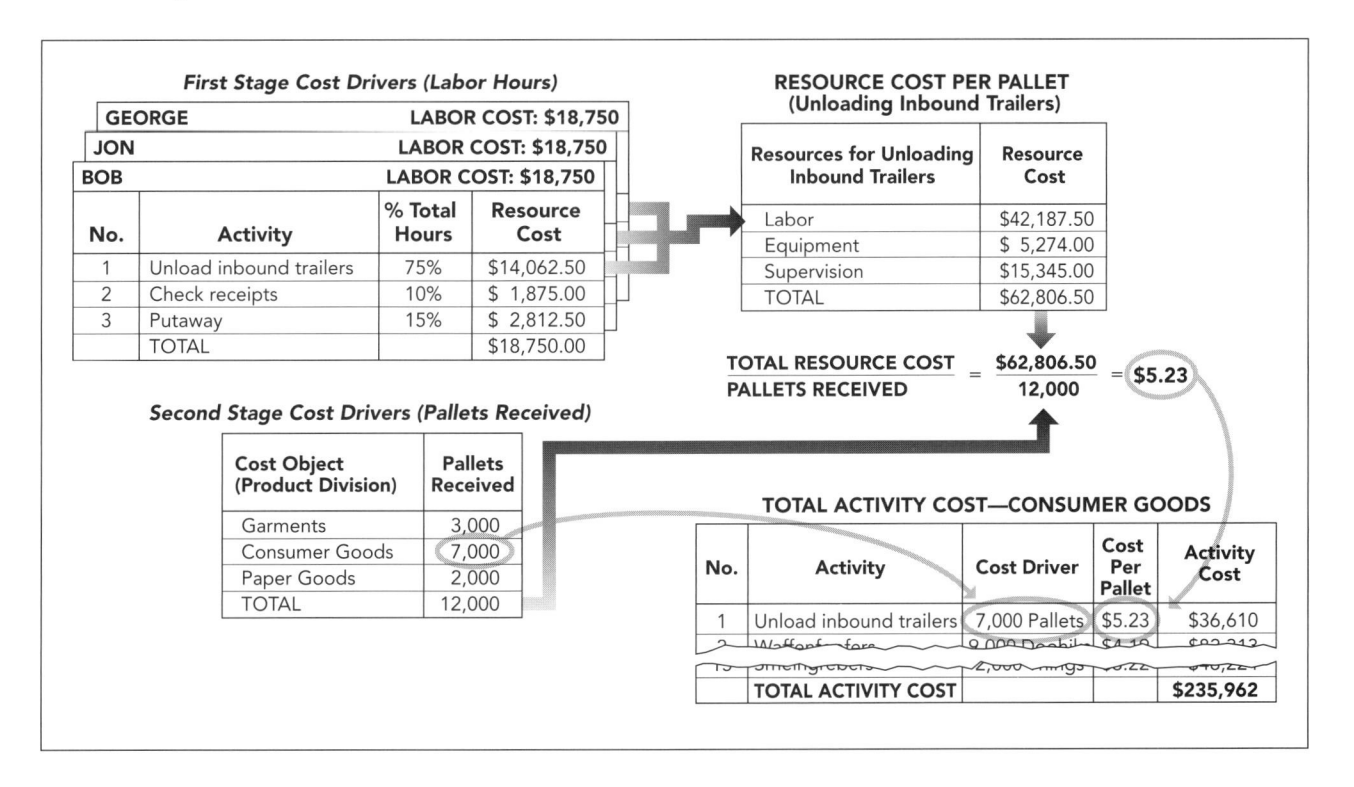

*First Stage Cost Drivers (Labor Hours)*

| GEORGE | LABOR COST: $18,750 |
| JON | LABOR COST: $18,750 |
| BOB | LABOR COST: $18,750 |

| No. | Activity | % Total Hours | Resource Cost |
|---|---|---|---|
| 1 | Unload inbound trailers | 75% | $14,062.50 |
| 2 | Check receipts | 10% | $ 1,875.00 |
| 3 | Putaway | 15% | $ 2,812.50 |
|   | TOTAL |   | $18,750.00 |

*Second Stage Cost Drivers (Pallets Received)*

| Cost Object (Product Division) | Pallets Received |
|---|---|
| Garments | 3,000 |
| Consumer Goods | 7,000 |
| Paper Goods | 2,000 |
| TOTAL | 12,000 |

**RESOURCE COST PER PALLET**
**(Unloading Inbound Trailers)**

| Resources for Unloading Inbound Trailers | Resource Cost |
|---|---|
| Labor | $42,187.50 |
| Equipment | $ 5,274.00 |
| Supervision | $15,345.00 |
| TOTAL | $62,806.50 |

$$\frac{\text{TOTAL RESOURCE COST}}{\text{PALLETS RECEIVED}} = \frac{\$62,806.50}{12,000} = \$5.23$$

**TOTAL ACTIVITY COST—CONSUMER GOODS**

| No. | Activity | Cost Driver | Cost Per Pallet | Activity Cost |
|---|---|---|---|---|
| 1 | Unload inbound trailers | 7,000 Pallets | $5.23 | $36,610 |
|   | TOTAL ACTIVITY COST |   |   | $235,962 |

To calculate the activity costs per cost driver, the total resources used by an activity is divided by the throughput experienced by a cost object. In Figure 59, the total resource costs of unloading inbound trailers is divided by the total number of pallets received. Resource costs for labor and overhead totaled $62,806 and 12,000 pallets were received. The activity cost per pallet is then $5.23 ($62,806 ÷ 12,000). The cost model assigns different cost drivers to different activities. Picking orders might be driven by the number of lines; shipping by the number of shipments.

The chart in Figure 60 is a sample worksheet for developing an ABC model of a product using traditional warehousing methods and the expected savings if included in a pre-allocated supplier consolidation program. Here the supplier provided the added cost of preparing cross dockable pallets. Activities that remain unaffected, whether one is cross docking or not, have been omitted. For this product, costs for receiving and shipping would remain with cross docking, but zero costs would be assigned for inspection, storage and staging.

**FIGURE 60**

**SKU Comparative Cost Model/Activity-Based Costing Method**

SKU Number/s: _____    Length: _____    Avg Inv: _____

SKU Description/s: _____    Width: _____

Supplier: _____    Height: _____    Inv. Turns: _____

Special Requirements: _____    Weight: _____

| | | Traditional Warehousing | | Cross Docking | |
|---|---|---|---|---|---|
| ACTIVITY | Annual Volume of Cost Drivers A | Cost Per Cost Driver B | Total Costs A × B | Cost Per Cost Driver C | Total Costs A × C |
| **Supplier Costs** | | | | | |
| Prepare cross dock pallets | | | | | |
| **Cross Dock Operator Costs** | | | | | |
| Receiving/Unloading (example) | 7,000 pallets | $5.23 | $36,610 | $3.54 | $24,780 |
| Inspection | | | | N/A | $0 |
| Putaway | | | | N/A | $0 |
| Storage | | | | N/A | $0 |
| Staging | | | | | |
| Shipping | | | | | |
| **Customer (Retail Store) Costs** | | | | | |
| Receiving/Unloading | | | | | |
| Storage (backroom) | | | | | |
| Stocking | | | | | |
| Retail Occupancy (cube) | | | | | |
| **ANNUAL COSTS PER PRODUCT** | | | D | | E |
| **SYSTEM SAVINGS WITH CROSS DOCKING** | | | | | D – E |

All cost figures should be rationally allocated or assigned to an individual product.
Activities such as shipment to stores which are not affected in either system have not been included.
Supplier costs provided—no ABC analysis

Ideally all of the members of the supply chain should incorporate their ABC analysis into one aggregate model. However, this may be expecting too much because ABC is still not used widely. A survey conducted by Ohio State University of 100 firms showed that only 10 percent of the companies use ABC in logistics. Reasons for not using ABC included: "higher management priorities; ABC is too complicated; not cost justified; other approaches appear more practical; and not sure how to proceed with implementation."[5] Therefore, the use of ABC may be limited to activities performed within the CDO's facility. Additional costs incurred by the supplier and or customer, based on their own analysis, are calculated and added to the ABC analysis of the CDO.

Either DPP or ABC can be used to determine savings and costs across the supply chain. With profitability and savings clearly stated in an accurate tabulation of costs, the cross dock team places itself in a better bargaining position with qualified parties involved in the cross dock effort.

## 4.3  Step 3: Calculate the exact costs and savings

With a breakdown of specific costs, the final determination of exact cost and savings to be shared can be determined. This is only true if negotiations have been made in Phase 1, Step 6 to share the actual monetary costs and savings of a cross dock program. In some cases, suppliers increase the acquisition or purchase price of their product so that their costs can be shared with the CDO.[6] Other CDOs provide a cash-back dollar amount to the supplier to share the savings gained from the program. Final policies that supplier and CDO agree upon should be determined and documented.

## Notes

1. For more information on this and other elements of costs, refer to Speh, Thomas W., Ph.D., *A model for determining total warehousing costs: For private, public and contract warehouses*, Illinois: Warehousing Education and Research Council, 1990.
2. Novack, Robert A., et al. *Creating Logistics Value: Themes for the Future*, Illinois: Council of Logistics Management, 1995, p. 152.
3. Martin, Christopher, "Integrating Logistics Strategy in the Corporate Financial Plan," *The Logistics Handbook*, New York: The Free Press, 1994, p. 256.
4. Haedicke, Jack, "The ABCs of Profitability," *Progressive Grocer*, January 1994, p. 38.
5. Pohlen, Terrance, "Applications of Activity-Based Costing Within Logistics: Who is Using Activity-Based Costing and Where?" *Annual Conference Proceedings*, Council of Logistics Management, Oct. 16–19, 1994.
6. As shown in our example in Section 4.2.

# Phase 4: Implementing and Maintaining the Cross Docking System

**Objective of Chapter 5:** *To provide a step-by-step implementation plan of the proposed design.*

| | |
|---|---|
| **5.1** | Step 1: Form a cross-functional team and verify objectives |
| **5.2** | Step 2: Locate cross dock center and select site (only for new facilities) |
| **5.3** | Step 3: Develop an implementation schedule |
| **5.4** | Step 4: Detail plans of required changes |
| **5.5** | Step 5: Train personnel |
| **5.6** | Step 6: Procure equipment |
| **5.7** | Step 7: Procure or upgrade information systems |
| **5.8** | Step 8: Prepare cross dock site |
| **5.9** | Step 9: Implement a pilot program |
| **5.10** | Step 10: Implement cross docking on a system-wide basis |
| **5.11** | Step 11: Periodically review cross docking operation |
| **5.12** | Step 12: Consider improving and expanding the cross dock program |
| **5.13** | Pitfalls to avoid and lessons learned |

# CHAPTER 5

# Phase 4: Implementing and Maintaining the Cross Docking System

The moment that first pallet leaves the supplier, irreversible events are set into motion. Everyone involved can only hope that the whole network of product, trading partners, equipment, and information is situated at the correct place and time so that each part of an effective and efficient cross docking operation can be carried out. For this to happen, a well thought-out implementation plan is a necessity.

The best plan is doomed to failure if not implemented effectively. Many barriers to implementation revolve around resistance to change. More often than not, cross docking, which impacts so many different people from varying levels of the supply chain, is bound to hit some resistance. Causes for such resistance stem from uncertainty about the new strategy, failure to see the need for the proposed change, fear of obsolescence, loss of jobs, poor approach by the design planners, or inappropriate timing—to name a few. To minimize this resistance, the need for cross docking should be explained thoroughly and convincingly. Major changes should be introduced in stages at the most appropriate time.[1] Developing a detailed implementation plan is a prerequisite to effective change management.

The following steps to implementation should be used *as a guide*. Some of these steps consist of specific activities that may occur simultaneously with other activities and *may not be in the order indicated*. Because cross docking can be as simple as moving full pallets within an existing facility or as complex as the installation of an automated sortation system in

**FIGURE 61**

**Phase 4: Implementation**

| | |
|---|---|
| STEP 1 | Form cross-functional team and verify objectives |
| STEP 2 | Locate cross dock center and select site (Only for new facilities) |
| STEP 3 | Develop an implementation schedule |
| STEP 4 | Detail plans of required changes |
| STEP 5 | Train personnel |
| STEP 6 | Procure equipment |
| STEP 7 | Procure or upgrade information systems |
| STEP 8 | Prepare cross dock site |
| STEP 9 | Implement a pilot program |
| STEP 10 | Implement cross docking system-wide |
| STEP 11 | Periodically review cross docking operations |
| STEP 12 | Consider improving and expanding the cross dock program |

a new facility, the steps in the process can vary significantly and should be approached on a case-by-case basis. Figure 61 summarizes the process.

## 5.1 Step 1: Form a cross-functional team and verify objectives

### Forming the Team

Implementation must be carried out systematically by the appropriate people who have been involved in planning and selecting the final cross docking solution and who have the ability and time to coordinate and manage diverse activities to make the design a reality. The implementation team should be cross-functional and gain support from all the affected parties. In one implementation, the accounting group, unaware of the program, held product for the mandatory three-day credit approval before releasing it to the warehouse. Because of this hold, the product could not be moved and was placed in temporary storage, negating all the benefits of cross docking.

In the planning process, managers for operations, facilities, information systems, transportation, human resources, and order processing must work together as a team and understand each other's role in the program. Representatives from the suppliers should be involved early in the process. Cross docking relies on suppliers' support and involvement during planning and implementation. Representatives from the receiving end of the pipeline must also be involved periodically. Production managers and store managers will need to be aware of revised delivery schedules. In return, they need to supply production and merchandising plans so that the CDO can anticipate fluctuations in volume. In third party environments, an engineer is typically needed for logistics design, and for costing of the new strategy. When a new facility specific for cross docking is required, senior management will need to be involved in the facility location and site selection process. The main responsibility of the implementation team is to plan and execute the cross docking program. Once under way, only a few key members of the team are charged with monitoring and strengthening the system.

### Using Consultants

Implementation must be transparent to the customer. Thus, people involved in implementation are usually expected to perform their regular functions while preparing the move towards cross docking. For some companies, this is not easy, and using consultants is often considered. While consultants may not know the business as well as the company's own employees, they keep the project moving forward and help anticipate any problems. Consultants can also bring expertise and experience from other cross docking implementation programs. They may also provide the following specific tasks on an as-needed basis:

- Selecting and implementing a Warehouse Management System
- Evaluating applicability of existing facilities
- Evaluating facility construction proposals
- Developing a move-in plan
- Training facility management staff during the start-up and debugging of a new warehouse operation
- Providing assistance in terms of staffing and job placement
- Bringing impartiality in any decision-making process

**Verifying Objectives**

Objectives, which were clearly stated in the first phase of the planning process, should again be verified and communicated to all members of the implementation team—especially those who have not been involved previously. This allows team members to work towards the same objectives and to agree on the steps to be taken.

## 5.2 Step 2: Locate cross dock center and select site (only for new facilities)

Where a new facility will be constructed specifically for cross docking, location and site selection are necessary. Senior management often charge the core members of the cross dock study team with determining the suitability of the site.

*Quantitative location models* are used to mathematically determine the best geographical area to locate the center. Computer models are especially useful for determining the optimum cross dock center location(s) within a network of multiple facilities of suppliers and customers. However, there are cases when the location of a facility may simply be a matter of situating it close to the point of demand. For example, cross docking centers in the automotive industry are located within a 24-hour service period of assembly plants requiring the just-in-time delivery of component parts.

After the general geographical area is determined through location models, *site selection models* provide a more qualitative tool for selecting the appropriate community and specific site for the new facility. Key factors in selecting a site include labor availability, land costs, standard of living, government incentives, and transportation

CHAPTER 5   *Phase 4: Implementing and Maintaining the Cross Docking System*

access. A rating system should be used to select the most feasible site and then the company may decide to purchase the land and construct a building to suit its needs, to purchase a pre-built facility, or to lease a pre-built facility.[2]

## 5.3   Step 3: Develop an implementation schedule

With the site determined, the team can put together a preliminary implementation schedule that may include the following activities:

- Develop a summary GANTT chart project schedule listing the major elements of the implementation (typically 60 to 100 steps).[3] Each step will include:
  — Person responsible
  — Duration
  — Dependencies (steps that must be completed or at a certain level of progress before the next step can begin)
  — Slack time (if zero, the step is on the *critical path*)

**FIGURE 62**

Sample Project Implementation GANTT Chart for Cross Docking

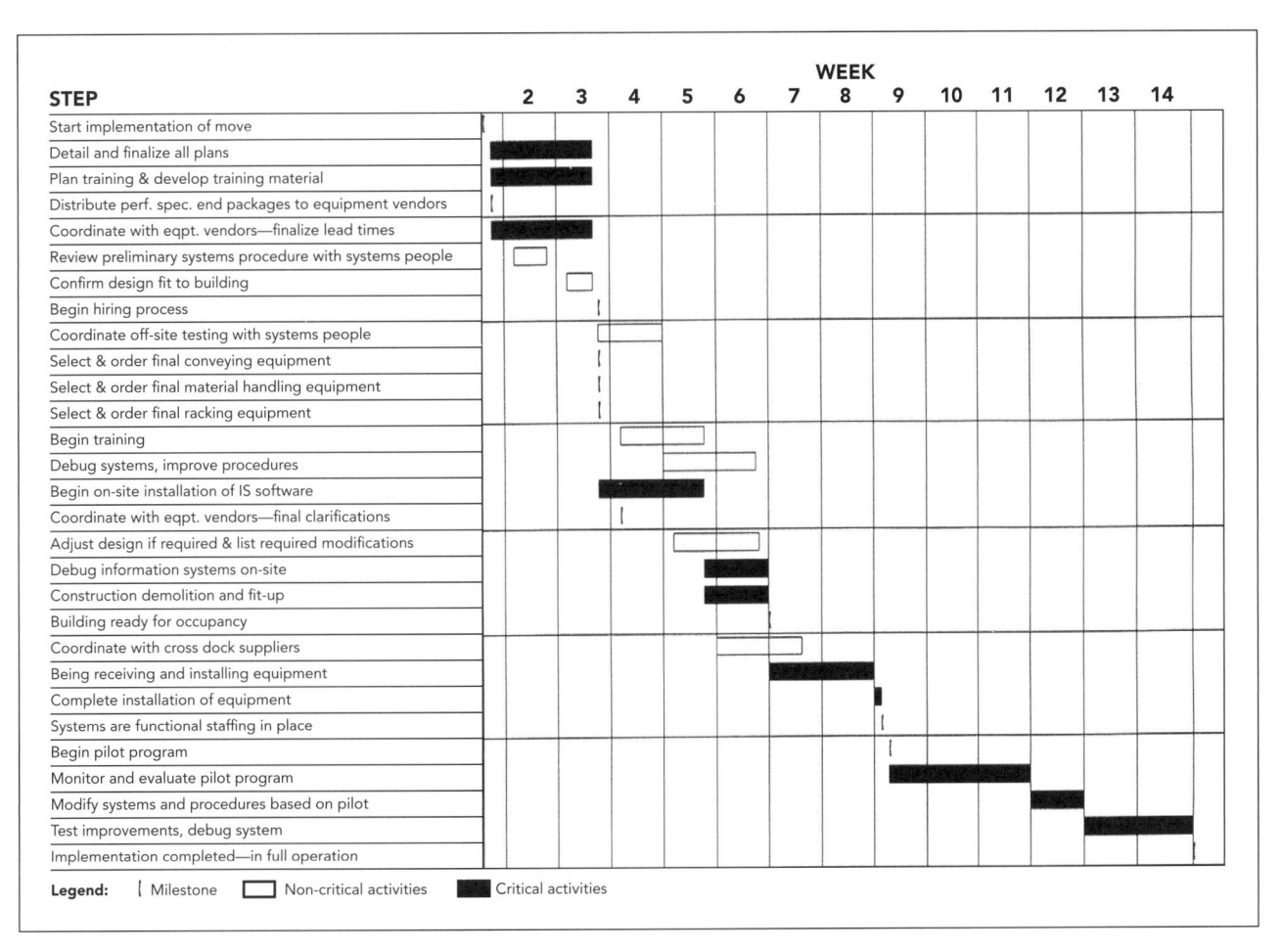

| STEP | 2 | 3 | 4 | 5 | 6 | 7 | 8 | 9 | 10 | 11 | 12 | 13 | 14 |
|---|---|---|---|---|---|---|---|---|---|---|---|---|---|
| Start implementation of move | | | | | | | | | | | | | |
| Detail and finalize all plans | | | | | | | | | | | | | |
| Plan training & develop training material | | | | | | | | | | | | | |
| Distribute perf. spec. end packages to equipment vendors | | | | | | | | | | | | | |
| Coordinate with eqpt. vendors—finalize lead times | | | | | | | | | | | | | |
| Review preliminary systems procedure with systems people | | | | | | | | | | | | | |
| Confirm design fit to building | | | | | | | | | | | | | |
| Begin hiring process | | | | | | | | | | | | | |
| Coordinate off-site testing with systems people | | | | | | | | | | | | | |
| Select & order final conveying equipment | | | | | | | | | | | | | |
| Select & order final material handling equipment | | | | | | | | | | | | | |
| Select & order final racking equipment | | | | | | | | | | | | | |
| Begin training | | | | | | | | | | | | | |
| Debug systems, improve procedures | | | | | | | | | | | | | |
| Begin on-site installation of IS software | | | | | | | | | | | | | |
| Coordinate with eqpt. vendors—final clarifications | | | | | | | | | | | | | |
| Adjust design if required & list required modifications | | | | | | | | | | | | | |
| Debug information systems on-site | | | | | | | | | | | | | |
| Construction demolition and fit-up | | | | | | | | | | | | | |
| Building ready for occupancy | | | | | | | | | | | | | |
| Coordinate with cross dock suppliers | | | | | | | | | | | | | |
| Being receiving and installing equipment | | | | | | | | | | | | | |
| Complete installation of equipment | | | | | | | | | | | | | |
| Systems are functional staffing in place | | | | | | | | | | | | | |
| Begin pilot program | | | | | | | | | | | | | |
| Monitor and evaluate pilot program | | | | | | | | | | | | | |
| Modify systems and procedures based on pilot | | | | | | | | | | | | | |
| Test improvements, debug system | | | | | | | | | | | | | |
| Implementation completed—in full operation | | | | | | | | | | | | | |

**Legend:**   | Milestone     ☐ Non-critical activities     ■ Critical activities

- Maintain and update the implementation schedule as the project progresses
- Track delays or changes in elements that may threaten the completion date and evaluate changes in other steps to keep the project on schedule
- Periodically distribute copies of the updated schedule to appropriate team members (or update the password-protected project schedule web site)

Figure 62 illustrates a sample Gantt chart for a cross docking implementation in an existing facility. Actual delivery and installation of equipment may take longer than originally anticipated. Maintaining the schedule on the computer allows the team to update the chart and quickly anticipate the effects of schedule revisions for future completion dates.

## 5.4  Step 4: Detail plans of required changes

Plans for the required changes should be finalized and described in detail for operations, facility, equipment, information systems, and transportation (Figure 63). These plans should include final operating specifications for new equipment, layouts, specific facility upgrades, etc. For suppliers who will be required to transform their product to suit the cross docking operation, specifications should document agreed-upon details (from Phase 3) of how the product must be presented to the cross docking center. Because the predictable can be unpredictable, contingency plans must be formulated to avoid unplanned breakdowns.

### Operational Plans

Under the operations manager's supervision, procedures must be described in detail for every aspect of the operation. Process charts from order processing to shipping should standardize the movement of product, people, equipment, and information (Figure 64). Frequently, products follow multiple paths through the facility. The flow path of full pallets will differ from the flow path of full cartons that require sortation. Each individual flow path must be defined and documented. Address the merging of cross docked SKUs with SKUs picked from inventory and prioritize cross docked items. Refer to Figure 65 for a sample checklist of elements needed in an operational plan.

### Facility Plans

Under the facilities manager's supervision, anticipated changes in the layout need to be clearly indicated. When existing operations

**FIGURE 63**

**Plans for Proposed Cross Docking System**

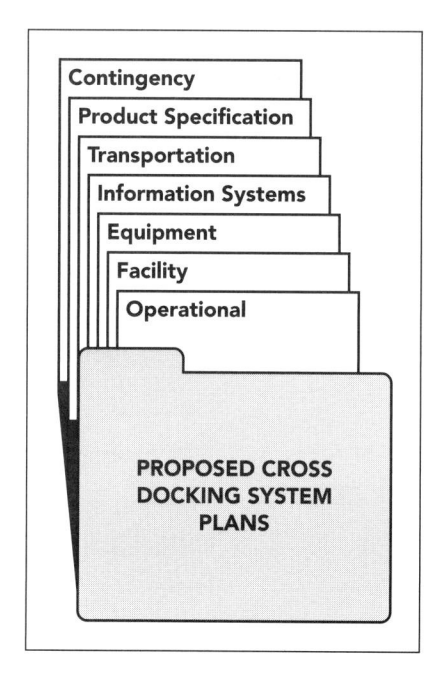

# CHAPTER 5 Phase 4: Implementing and Maintaining the Cross Docking System

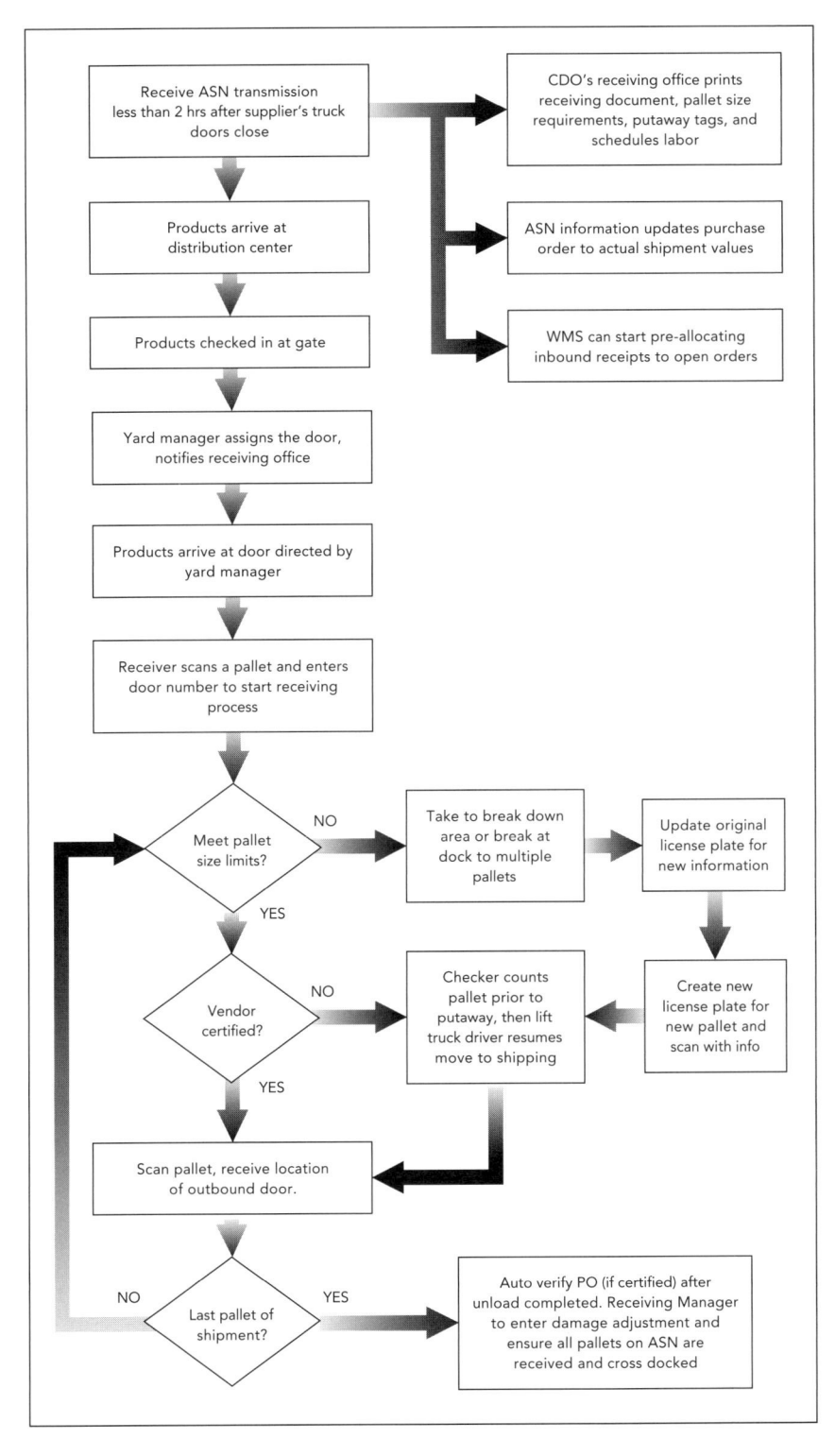

**FIGURE 64**

**Sample Process Flow Chart for Receiving**

Adapted from "Cross Docking: Process for Success," *Special Report for FMI/GMA Replenishment Excellence Conference.* New Orleans, LA, September 1995.

**FIGURE 65**

**Sample Operational Plan Checklist**

> **Have process charts been developed for each of the following areas? (Appropriate equipment and personnel for each action should be indicated.)**
> ☐ Receiving (cross docked items and combination cross docked items and storage)
> ☐ Inspection (if necessary)
> ☐ Labeling (if required)
> ☐ Sortation and consolidation of orders (if required)
> ☐ Merging of cross docked SKUs with SKUs picked from inventory
> ☐ Shipping
> ☐ Yard management
>
> **Have the following flows been established based on the proposed layout and information system?**
> ☐ Product flow
> ☐ People flow
> ☐ Equipment flow
> ☐ Information flow
> ☐ Transportation flow (for revised networks)

will be renovated, facility plans must adopt a phased-in approach to keep business running as usual. The exact number and location of pallet racks to be removed, if any, should be clearly indicated, and the quantity and arrangement of installing equipment, such as pallet racks and conveyors, should be finalized. These plans will be

**FIGURE 66**

Sample Facility Plan Checklist

> **If retrofitting an existing facility, have phased-in facility plans been developed? Do they indicate the following:**
> ☐ Exact number and location of pallet racks to be removed, if any
> ☐ Built-in impediments to operational flows that CAN be relocated such as walls, doors, etc.
> ☐ Built-in impediments to operational flows that CANNOT be relocated such as HVAC systems, sprinklers, electrical, etc.
> ☐ Changes to the yard to accommodate increased throughput volume, if any
> ☐ Approval by facility manager of the required changes
>
> **Do facility layouts for the proposed cross docking system indicate the following:**
> ☐ Final arrangement of installed equipment such as pallets, racks and conveyors
> ☐ Proper aisle requirements for personnel and equipment
> ☐ Final arrangement of workstations
> ☐ Dock door assignments
> ☐ Power and other utility requirements
> ☐ Location of stationary terminals

**FIGURE 67**

Sample Equipment Plan Checklist

> **For each type of equipment to be purchased, the following specifications should be indicated:**
> ☐ Description of operation for which the equipment will be used
> ☐ Dimensions and weights of the product to be handled
> ☐ Bar code labeling requirements, if any
> ☐ Required throughput rates and capacities
> ☐ Quantity required
> ☐ Any exceptions
> ☐ Any supporting information and design analysis such as layouts and flow diagrams
> ☐ Limits on time (where expediency is necessary)
>
> **Have equipment specifications been reviewed by facility and equipment manager to verify capacities, clearance limits, and compatibility with existing equipment?**

used by the builders, contractors, and systems people in making the actual modifications to the facility, installing equipment, adding power where appropriate, and handling other utility issues. Refer to Figure 66 for a sample checklist of elements needed in a facility plan.

## Equipment Plans

Under the operations manager's supervision, equipment plans should indicate the type of equipment to be purchased, the required throughput for that equipment, the minimum and maximum dimensions and weights of the material for which the equipment is used, along with other operating specifications. These specifications will be provided to vendors for bidding. The facility's maintenance supervisor will need to review equipment plans and verify capacities and compatibility with existing equipment. Refer to Figure 67 for a sample checklist of elements needed in an equipment plan.

## Information Systems (IS) Plans

Under the supervision of the information systems manager, IS plans should indicate detailed software changes to the WMS. Hardware requirements and operating specifications should be finalized and documented. Compatibility with existing equipment should be verified. Most significantly, the operations manager should communicate to the systems manager what is expected from the Warehouse Management System.

The information systems manager must document how cross docking will impact other levels within the company and communicate those changes to appropriate personnel. For example, every carton scanned at receiving may automatically begin the processing for paying an invoice. The accounting department must be made aware of this change. Suppliers often consider quicker payment for their merchandise as a big plus as their customers move towards cross docking. External to the warehouse, a system of communication among suppliers, CDOs, and customers should be established. IS-related methods, level of standardization, and schedule of data and information transfer should be well documented and agreed upon by suppliers and customers. Refer to Figure 68 for a sample checklist of elements needed in an information systems plan.

**FIGURE 68**

**Information Systems Plan Checklist**

> **Have detailed changes to the WMS been documented and approved for the following areas to enable cross docking?**
> ☐ Receiving (cross docked items and combination cross docked items and storage)
> ☐ Inspection (if necessary)
> ☐ Labeling (if required)
> ☐ Sortation and consolidation of orders (if required)
> ☐ Merging of cross docked SKUs with SKUs picked from inventory
> ☐ Shipping
> ☐ Yard management
>
> **Have hardware requirements and their operating specifications been documented?**
>
> **Has compatibility with existing equipment been verified?**
>
> **Have inter-departmental changes to the WMS due to cross docking (if any) been documented and disseminated to company personnel beyond the cross dock center?**
>
> **Has a standard system of communication and information exchange among suppliers, CDOs, and customers participating in the program been documented and disseminated?**

## Transportation Plans

Under the transportation manager's supervision, any scheduling or routing changes to stores or production plants should be indicated and communicated to key trading partners. Stores or assembly plants will need to know if delivery should be increased from a once a week to a twice a week schedule. If the

**FIGURE 69**

**Transportation Plan Checklist**

> **Have scheduling and routing changes been documented, approved, and disseminated to key trading partners?**
>
> **Have the following transportation-related activities for cross docked items been documented along with software changes to the Transportation Management System?**
> ☐ Prioritization protocol
> ☐ Yard management
> ☐ Loading and unloading procedures
> ☐ Dock door scheduling and appointments
> ☐ Any other activity that may affect transportation and warehousing operations

change is from warehouse bypass (such as direct store delivery) to cross docking, the effect on transportation can be dramatic. More suppliers will converge at the CDO's receiving docks while less activity will be anticipated at each store. With either method, truck movement in the yard in front of the receiving and shipping docks must be regulated. Yard management programs should have the ability to schedule dock appointments to reduce congestion and prioritize the receiving and shipping of cross docked items. Transportation management software should still allow the DC to save shipping costs by performing routines that allow building economical loads and automatic selection of the best carrier for the cross docked shipment based on cost and service. A switch to a door-per-store concept should be documented. Refer to Figure 69 for a sample checklist of elements needed in a transportation plan.

**Product Specification Plans**

This is one of the more critical plans to establish between the supplier and the CDO if the CDO expects the supplier to ship product differently from its current form. A supplier who is required to label products or build store-specific pallets needs this plan so that he is aware of CDO specifications, equipment restrictions and tolerance limits. For example, product to be cross docked on a fully-automated pallet conveyor system cannot exceed specific dimensions before entering the system. The receiving function will be severely compromised if a unit load is too tall and requires restacking. When cartons are to be scanned by automated systems, the correct placement of the bar code label is critical to minimize read errors, and the plan should specify the exact placement of bar code labels. In preallocated supplier consolidation, an assortment of SKUs needs to be consolidated by the supplier. The CDO may want the SKUs arranged to mirror store layouts to gain restocking efficiencies at the stores. If cross dock centers supply production, component parts

**FIGURE 70**

**Product Specification Plan Checklist**

> **Have the following specifications for the cross docked product been documented, approved, and disseminated to the supplier and other trading partners?**
> ☐ Logic of desired SKU configuration on pallet or in carton or tote (such as mirroring store layouts)
> ☐ Minimum and maximum dimensions and weights of pallets
> ☐ Minimum and maximum dimensions and weights of cartons or totes
> ☐ Proper placement of bar code labels
> ☐ Any exceptions
> Note: This plan is not necessary if the CDO does not expect the supplier to ship product any differently from its current form.

may need to be shipped together for a production run. This should also be specified in this document. Refer to Figure 70 for a sample checklist of elements needed in a product specification plan.

## Contingency Plans

Trucks may run late. Product may arrive incorrectly or damaged. Contingency plans help maintain control and prevent system breakdown. One method for ensuring that unplanned events are immediately communicated to the parties involved would be to establish simple, standard procedures. For example, through EDI the CDO tells drivers when and where to pick up and deliver. Once the task is completed, a reply must be sent back to confirm completion by the trucking company. If a reply is not received, contingency plans can be put into effect. The most efficient cross docking in the world is useless if products sit on the loading dock waiting for a truck to pick up. To avoid problems, Miller Brewing Co. uses *converting*—giving a load away to a truck on hand when one truck doesn't show up on time.[4]

If product quantity runs short, maintaining a half a week's supply of inventory will reduce stock-out conditions at the stores. If no

**FIGURE 71**

**Contingency Plan Checklist**

> **Have control procedures been documented so that a break from the procedure launches contingency plans without delay?**
>
> **Have contingency plans been documented, approved, and disseminated for the following unexpected events?**
> ☐ Inbound tractor trailer delay
> ☐ Outbound tractor trailer delay
> ☐ Incorrect and/or damaged product
> ☐ Excess or shortage of product
> ☐ Automated equipment breakdown

inventory is available, a formula for reducing the product quantity delivered to each store must be agreed to in advance. If product is in excess, a procedure must be established to return the overage to the supplier. If excess product needs to be stored, the supplier and the CDO must agree on a storage fee to be charged back to the supplier. Contingency plans should be agreed upon, carefully documented, and distributed to all the parties involved before any cross docking can begin. A "hold-over" charge is usually agreed upon for products that are unexpectedly left on a dock for a longer period of time, but in reality this rarely happens. Refer to Figure 71 for a sample checklist of elements needed in a contingency plan.

## 5.5 Step 5: Train personnel

Once plans are documented, preparing training programs can begin. Training specific for cross docking is recommended for personnel responsible for receiving, sorting, and shipping the cross docked products. Training costs may be lower than in traditional warehousing because this strategy involves fewer handling activities. In addition to the dedicated pool of cross docking workers, other warehouse personnel should be acquainted with the basic procedures because additional labor may be required during peak periods. Outside the warehouse, drivers must be trained to be available to move trailers to meet the unloading/loading schedule. Training activities may include the following:

- Planning training sessions for different cross docking functions
- Developing training materials using operation plans developed in Step 4
- Starting the hiring process (if required)
- Training personnel
- Inviting supplier and customer personnel to training sessions so that they can be more aware of the operations
- Monitoring newly trained personnel in the pilot program
- Updating training sessions to reflect gaps in training and performing
- Communicating gaps to newly trained personnel

## 5.6 Step 6: Procure equipment

Any new equipment required for cross docking must to be purchased. Plans developed in Step 4, and a schedule for equipment installation and use, are part of a *performance specification bid* package, resulting from the design phase (Phase 2), to be distributed to qualified vendors. Other activities related to the procurement of the equipment include:

- Distributing *performance specification bid* packages for qualified vendors
- Coordinating with vendors to answer equipment questions
- Receiving quotes for equipment
- Evaluating quotes
- Selecting vendors and equipment
- Confirming delivery dates
- Adjusting design and budgets based on equipment vendor specifications and costs
- Receiving and installing equipment
- Monitoring equipment installation to confirm that it conforms to the requirements
- Testing and monitoring equipment in pilot program
- Fixing gaps or snags in equipment with vendor support

Equipment for a cross docking system may include material handling equipment, such as lift trucks, electric pallet jacks and pallet and carton conveyors and storage equipment such as pallet racks, flow racks or carousels.

There is no standard sequence in installation. It may differ greatly depending on a number of factors:

- *Time.* Conveyor systems take longer than pallet racks so there is generally pressure to start them earlier.
- *Structural dependencies.* Mezzanines have to be built before shelving on the mezzanine can be installed.
- *Condition of the building.* Pallet racks and mezzanines do not need a totally sealed, secure building with full electric, while a complex conveyor sortation system might. On the other hand, the pallet racks or mezzanines might hinder access to the ceiling, so the lights and sprinklers must be completed first.
- *Union issues*
- *Equipment specification issues.* It is usually appropriate to install pallet racks before the wire guidance system, because the wire is required to be in the exact middle of the rack aisles. The wire guidance may be installed first, but the racks will have to go in *exactly* as indicated in the plan.

## 5.7 Step 7: Procure or upgrade information systems

Information systems are typically the heart of a cross docking operation. *Information must travel with the product* for warehouse personnel to know each item's next destination. If the operation is complex, a completely new Warehouse Management System may be required. Initial cross docking projects may only require an upgrade or modification of the present WMS.

Cross docking users interviewed for this research expressed their frustration over this step in the implementation. There are frequent delays caused by:

- Systems "not talking the same language."
- Costs that are usually higher than anticipated.
- Time for programming and testing that is longer than anticipated.
- Technology that is developing so rapidly that it is difficult to keep up with changes and upgrades.
- Operational details and logic were overlooked.
- The complexity factor of certain modules of a system.

**REAL WORLD EXAMPLE**

**Information Systems Glitch
Slows Down Ford's New Mixing Centers**

The implementation of Norfolk Southern's (NS) Ford automotive mixing centers encountered a minor glitch in January 1998. These four mixing centers—essentially cross dock centers using rail transportation and located in Kansas City; Chicago; Fostoria, OH; and Shelbyville, KY—unload vehicles from 21 Ford North American assembly plants, sort them by destination, and load them back onto railcars with other models bound for the same destination. (See sidebar in the introductory section of Chapter 1 for a full description.) However, problems with the information systems that tracked a fleet of 15,000 cars from assembly line to dealer clogged the centers with vehicles and railcars. Mixing center operators were not getting the information needed to send vehicles on railcars on their way. Some railcars had to be diverted back to the assembly plants, while nearby yards and other railroads helped ease the congestion out of the centers. NS management insisted that the problem did not reflect a fundamental flaw in the mixing center concept. By early March, the software glitch was fixed and the system was running nearly up to speed. Other factors contributed to the problem: the need to break in new contractor crews and a backlog with a haulaway trucking company. Source: Stephens, Bill, "Roads smoothes out for NS auto hubs," *Trains*, June 1998, pp 21-22.

Companies should start with one technology piece at a time so as not to overwhelm the team. Specific activities to upgrade existing information systems may include:

CHAPTER 5   Phase 4: Implementing and Maintaining the Cross Docking System          137

- Providing information system specifications developed in Step 4 to programmer
- Coordinating with programmers and answering any IS-related questions
- Reviewing preliminary user interface design, software process design, database design with the appropriate personnel
- Testing off site
- Monitoring off-site tests
- Revising as required
- Installing systems on site
- Executing on-site systems test
- Monitoring/debugging on site and revising as necessary
- Declaring systems live and functional
- Executing systems in pilot program
- Monitoring systems in pilot program
- Revising as required
- Certifying that the systems upgrade is accurate and complete

## 5.8   Step 8: Prepare cross dock site

While information systems and equipment are acquired, the site for the cross docking operation should be prepared based on facility plans developed in Step 4. Preparation of a site may be as complex as building a new facility or, if retrofitting an existing one, could be as simple as removing a few pallet racks or adding a dock door. Activities related to the preparation of a cross dock site in an existing facility might include:

- Finalizing design drawings with builder/contractor modifications
- Planning construction and/or demolition without affecting existing operations
- Executing construction and/or demolition and fit-up (includes removal of storage racks for more operating dock space if necessary)
- Declaring cross docking area ready for installation

## 5.9   Step 9: Implement a pilot program

A pilot program is the full implementation of the new strategy on a much smaller scale. It may involve cross docking a few items from one or two of the most reliable suppliers over a set period of time. Launching a pilot program can:

- Determine how layout, equipment, personnel, information systems and other components can work efficiently and effectively to meet or exceed the design year throughput volumes.
- Evaluate how well warehouse employees understand their roles with regard to the new strategy, and detect any gaps in training.
- Allow the team to observe the integration of the cross docked products with products processed the traditional way at a manageable level.
- Allow preliminary measurement of costs and savings of the cross docking system and the comparison of these measures with industry standards or with actual cost objectives. Under the supervision of the operations manager, the following performance measures can be tracked at the CDO's facility:
  — Labor hours and labor cost per unit handled
  — Time from product receipt to product shipment
  — Labor utilization per shift
  — Order fill rate
  — Order accuracy rate
  — Level of inventory (if some inventory is maintained)
  — Compliance to CDO guidelines

For transportation, the following performance measurements can be tracked:
  — Time trailer spends at cross dock receiving and shipping
  — Cube utilization per shift
  — Rate for meeting scheduled deliveries (inbound and outbound)

From customers (stores or plants), the following measures should be tracked:
  — Number of claims and damages
  — Customer service observations
  — Inbound labor productivity

Employees should be encouraged to voice concerns and uncertainties before full implementation. Activities in a pilot program may include:

- Selecting products and suppliers with the highest cross docking ease factor (from Phase 1)
- Defining success criteria and milestones for the pilot

CHAPTER 5   *Phase 4: Implementing and Maintaining the Cross Docking System*     139

- Informing concerned suppliers, warehouse personnel, systems people, and customers of success criteria and milestones
- Confirming pilot starting date
- Implementing the pilot
- Monitoring and measuring performance
- Evaluating the results of the pilot program
- Modifying pilot as required
- Declaring pilot as fully functional

## 5.10  Step 10: Implement cross docking on a system-wide basis

The success of the pilot paves the way for the full implementation of the cross docking program. All of the equipment, systems, and staffing should be in place and fully functional. On-site support by the project team should continue after full implementation for a period of at least two demand cycles to ensure that the system is operating smoothly. Activities involved with the full implementation of a cross docking program in an existing facility may include the following:

- Declaring equipment and systems functional with staffing in place
- Receiving product from all selected suppliers
- Beginning the cross docking operation
- Starting shipping to customers
- Declaring system in full operation

## 5.11  Step 11: Periodically review cross docking operation

Once the system is declared fully operational, only a few key members of the implementation team need to execute the final two steps in the implementation process. The periodic review and possible expansion of the cross docking system can be carried out under the supervision of the operations manager with the assistance of supervisors involved with the entire process.

Periodic review of the cross docking operation allows the team to observe the effect of change to the system, to identify gaps in performance, and to encourage continuous enhancement to eliminate these gaps.

Everything changes. Demand for a product changes. Technology changes. The way people do business changes. Practitioners must monitor the operation to keep up with these changes. The same performance measures enumerated in the pilot should be continued in the full implementation of the program.

Supplier performance should be monitored continuously using the same criteria developed in the supplier review section of Phase 1. A decline in performance should trigger an investigation. In so doing, the company can be assured that the cross dock is operating at its best level. Periodic review also encourages continuous enhancements: Gaps in the systems can be addressed and recommendations provided to improve the system.

## 5.12 Step 12: Consider improving and expanding the cross docking program

Properly executed cross docking will provide considerable benefits and savings and compel a company to improve and expand the program. As players become more comfortable with the concept, additional products can be considered. The cost models discussed in Phase 3 of this process can be used to determine the cost-effectiveness of each new SKU for cross docking. These cost models will also aid in additional negotiations with old or new suppliers and customers regarding the sharing of benefits and costs. When adding products, however, the pilot program should be closely monitored. Limits on capacities may be breached, and new or higher state-of-the-art technology or automation may be required.

## 5.13 Pitfalls to avoid and lessons learned

The most important thing to remember is that cross docking alone is not the "Holy Grail" of warehousing. It can move product in and out of a facility, with no inventory, little handling, and in a shorter time period, but it requires an enormous support system and infrastructure to do so, such as:

- Suppliers must deliver the correct quantity of the correct product at the correct time
- Marketing has to plan demand and schedule promotions
- Accounting has to monitor costs and facilitate paying invoices
- Transportation has to move inbound and outbound loads efficiently and cost-effectively
- Warehouse personnel must move the product with its proper information
- Inventory management has to study demand and order the right amount of product at the right time so overstock and out-of-stock conditions do not occur

- A communications and systems network must be in place so that product moves based on the latest demand information

Following are other lessons learned from cross docking practitioners organized by phase.

## Phase 1: Assessment and Negotiation

- *Match logistics processes with strategic plans.* Promotional events must be planned and communicated to logistics so that the correct product can be cross docked in a timely manner. CDOs must work with suppliers to identify products in their marketing strategies that have the potential for cross docking.
- *Agree on specific terms beforehand.* After negotiations with suppliers and customers, document specific terms and distribute the document so that everyone can sign off on their agreement.
- *Focus first on products and suppliers that are easiest to cross dock.*
- *Work with suppliers who are willing to support a cross dock program.* Some suppliers may have the technology to support cross docking but may lack the willingness to negotiate savings and costs. CDOs have been known to drop suppliers who refuse to negotiate the added costs of creating custom pallets.
- *Pay attention to customer needs.* Certain customers may want their products to come consolidated in a particular way. Other customers may not have enough display area available for floor-ready pallets at their stores.
- *Consider using third party providers.* A supplier may be unwilling to create custom pallets that can be readily cross docked, thus the CDO may consider working with a third party provider who can supply these pallet customization and consolidation services. Careful analysis is required to ensure that costs for outsourcing are offset by benefits and savings from the program at the next point in the supply chain.
- *Emphasize internal collaboration and external cooperation.*
- *Establish open communication with all members of the supply chain.*

## Phase 2: Planning and Design

- *Build toward a higher level of technology one stage at a time.* Implementing cross docking is complicated enough. Simultaneously adding sophisticated, high-tech equipment at the start of the program can increase the complexity factor many times.
- *Build a new facility for cross docking with plenty of dock space.* Many facilities are being built with 60- to 70-foot dock depths. Richfood, a grocery retailer in Mechanicsville, VA, built a

state-of-the-art sortation facility primarily for cross docking, despite lagging behind in information systems technology. However, Richfood is confident that its new facility will enable them to cross dock through the new millennium.[5]

- *Observe other cross docking operations.* The best way to learn is by example. Observe how other companies are cross docking. Determine how certain aspects of their operation can work for you.

- *Ensure that your WMS can support cross docking.*

- *Ensure that external systems have some system of communication with your WMS.*

- *Standardize information.* Speaking the same language in the form of standardized data is key to smooth communication. Systems must be designed so that all parties at different levels can understand the data being relayed.

- *Emphasize the need for open communication and high product visibility.*

- *Do not get hung up on technology.* Consider the case of EDI. For many CDOs, having thousands of suppliers sending data electronically is unrealistic. Faxes are still a popular alternative. Two-dimensional bar code symbology can now place all EDI information on a bar code label. This single symbol contains hundreds of characters of data and can be read reliably with a hand-held scanner. It was identified in ANSI MH10.8.3, the ANSI standard on two-dimensional symbologies for unit loads and transport packages, as the way to transmit "paper EDI" among trading partners who wanted it but could not afford it.[6] With a single scan, all the information required to cross dock can be in the hands of the receiving person.

- *Make sure that the customer (store or plant) slated to receive product has the space available for it.*

- *Emphasize good dock design and safety.*

- *Consider simulation, especially when designing a complex fully automated, capital-intensive cross docking system.* Simulation allows the operation to be replicated on the computer and tested before any piece of equipment is purchased.[7]

## Phase 3: Justification and Cost Sharing

- *Establish true cost at the outset.* True savings from cross docking cannot be determined if a company does not maintain existing cost accurately. Knowledgeable consultants report that approximately three to six months are spent determining true baseline costs for clients who do not track them. Elements of costs and

CHAPTER 5  *Phase 4: Implementing and Maintaining the Cross Docking System*   143

actual cost figures can be determined from a general ledger or a financial statement of operations. Certain assumptions may need to be in place when converting costs to an item-by-item basis.

- *Build cost models that can be easily updated on the computer.* Tracking costs can be a tedious and labor- and time-intensive process. Building a template model on the computer allows one to test the cost-effectiveness of new products for cross docking, quickly update cost values when needed, or create cost-sharing scenarios for negotiation with suppliers.

- *View savings from a system-wide perspective and shared equitably between trading partners.*

- *Keep it simple.* Keep the model as simple as possible to make it easy for all of the parties involved to understand the costs and savings implications of cross docking.

- *Use the cost model to test other strategies.* With a cost model in place, the CDO might determine that strategies other than cross docking will be more cost efficient for a certain class of products. Tops Markets, a grocery retailer servicing over 200 stores uses an ABC cost model to determine whether a product should be cross docked or delivered directly to the store. Some products were determined to be more profitable if some suppliers shipped directly to the stores rather than having to go through their state-of-the-art facility.

- *Research building ABC as well as other cost models.* There are a number of references available for ABC and other cost accounting practices. Talk to companies who are using ABC. Research how companies justify capital investments.

## Phase 4: Implementation

- *Develop a master schedule on the computer.* The master implementation plan should be developed using a project planning software tool (such as Microsoft Project or Visio). There may be many updates and revisions to the schedule. Updating it on the computer enables quick and automatic rescheduling of future dependent activities and allows the project team to quickly determine any critical scheduling delays or shortfalls.

- *Document, document, document.* Product specifications, processes, and programming logic should be documented, communicated, and approved by the appropriate people. Grocery studies indicate that 54 percent of all receiving delays are caused by the need to restack pallet loads because suppliers do not build loads that suit the distributor's needs. A clear product specification plan documenting these needs and agreed upon by the supplier may eliminate this problem.[8]

- *Cross docking is not an exact science.* There are times that product will have to be brought in ahead of schedule to create a cost-efficient truckload. There are times when product must be held back based on constraints for efficient receiving and transportation. Generally a high percentage of product should flow through the warehouse without delay. The program should have the flexibility to adapt to unplanned changes.

- *Technology is only useful when applied properly.* There is no point in investing in expensive technology if proper procedures for its use are not followed. Many CDOs complain of receiving ASNs after the truck arrives. ASNs might provide a big advantage for cross docking, but they are useless if not sent on time.

- *Take slow steps initially and never compromise service levels.* The pilot program should have performance measures to track adherence to program requirements. Each step in the pilot should be invisible to the customer, thus enabling him to perform business as usual.

- *Once implemented, determine whether or not cross docking is paying off.* Develop a measurement system that will compare cross docking costs to storing product costs.

- *Plan for conditions that will negatively impact cross docking, instead of reacting.* Maintain a list of contingencies in case a link in the chain is broken or jeopardized.

Cross docking is a big undertaking involving personnel from different functions internal and external to the organization. It requires give and take with everyone involved with many details. The effort to assess, design, justify, and implement the program should never be underestimated. Systematic, proactive project management with proper planning and documentation will ensure that critical details are not overlooked, and a cross dock program can be implemented effectively and efficiently.

## Notes

1. Tompkins, et al. *loc. cit.*, p. 708.
2. For more information on facility location and site selection, see Chapter 3 of *Using Modeling to Solve Warehousing Problems*, Warehousing Education and Research Council, 1998. In addition, there have been a number of studies dealing with the modeling of cross dock networks. See Bibliography for details.
3. For more information on GANTT charts, refer to any operations research publication. Software programs such as Microsoft Project and Visio 2000 aid in developing GANTT charts on the computer.
4. Small, Stefanie, and McAndrews, Maureen, "Keeping Inventory Moving," *Warehouse Management*, September 1998, p. 35.

5 Garry, Michael, "A Blueprint of the Future," *Progressive Grocer*, October 1994, p. 63.
6 Moore, Bert, "Where do you want to know today," *Automatic I.D. News*, August 1999, p. 43. For more information, visit www.aiag.org.
7 For more information on simulation, refer to *Using Modeling to Solve Warehousing Problems*, Warehousing Education and Research Council, 1998.
8 Wagar, Kenneth, loc. cit., p. 188.

# CHAPTER 6
# A Case Study on Cross Docking

*This case study is adapted from a cross docking report completed by the staff of Gross & Associates. Names and relevant statistics have been changed for confidentiality. It illustrates a move from a traditional store-and-ship system to a predominantly post-allocated CDO consolidated cross docking operation without extensive spending on automation.*

**6.1** Background

**6.2** Generating cross dock system designs

**6.3** Economic analysis of alternatives

**6.4** Implementation

**6.5** One year of operation

**6.6** The future for S&O

# CHAPTER 6

# A Case Study on Cross Docking

## 6.1 Background

Saw & Order Inc. (S&O)[1] is a do-it-yourself hardware retailer with 83 stores in the Midwest and the Northeast. The company wants to reduce inventory investment, storage space, and cycle times in its present facility. Management recognizes that it can fulfill these objectives by implementing a cross docking system. Of the approximately 2,300 SKUs presently warehoused, they have identified 14 suppliers and 125 SKUs (approximately 5 percent) for immediate cross docking.[2] Predictable weekly sales patterns and high cubic movements are the principal criteria for selecting the appropriate SKUs to cross dock. Suppliers will continue filling approximately the same volume of products for the month. However, instead of receiving many trucks from the same supplier once a month, S&O may have trucks arriving at least once a week from a supplier. Suppliers are expected to continue to deliver the correct merchandise at the correct time and place. S&O will still be responsible for labeling the cartons after they arrive at the facility. However, the company is taking advantage of some of their suppliers' offers to build pre-assorted, "display ready," and store-specific pallets, which S&O wants to cross dock. If successful, S&O intends to increase the percentage of items to cross dock to over 30 percent of the total SKUs by the end of Year 3.

Data regarding present facility conditions and operations have been collected.[3] S&O currently stores its products in a traditional facility with a combination of bulk and single deep pallet rack storage modules. It uses conventional lift trucks and pallet jacks to put away, pick, and transport product within the facility. There are no conveyors or automated systems in use. Presently a picker fills orders for each store by using a pallet jack to pick cartons from the bottom levels of pallet racks. Warehouse personnel use radio frequency devices that process information in real time. There are no sortation systems in place and the company plans to stay in the facility for another 8 to 10 years. Although equipped with EDI and a "better-than-average" Warehouse Management System, the company

CHAPTER 6   A Case Study on Cross Docking

still considers itself to be in the early phases of transition to a more advanced information systems platform.

S&O would like cross docked products for specific stores to be delivered to the warehouse on the day of those stores' scheduled delivery. Ideally, the dock should be clear of any pre-allocated orders by the end of the day. Because it wants to achieve economies of scale from truckload shipments and because it does not have the truck fleet or the dock doors for a door-per-store scenario, some products may have to be received a day or two before a store's scheduled delivery. These items will to be staged in the processing area for a day or two before leaving the warehouse. To S&O, making room for temporary staging is deemed more cost effective than using less than truckload (LTL) shipments, expanding their truck fleet, or increasing the frequency of delivery to their stores. Currently, scheduled delivery to stores is once a week with 17 stores being serviced by the warehouse in one day. For now, S&O would like to maintain the same schedule. In addition, only the aggregate quantity of SKUs needed by the stores scheduled for delivery that day is received at the warehouse. As the company keeps growing, it intends to reassess these plans to maximize the benefits of cross docking—especially when planning a move to a new facility.

Based on this information, S&O embarked on the planning and design of a cross docking system.

## 6.2 Generating cross dock system designs

The first part of the design process generates specifications for different alternatives.[4]

### Step 1: Assemble the design parameters

Figure 72 summarizes the projected operations and facility design parameters for the new cross docking strategy. Year 1 shows the level of cross docking that S&O would like to achieve immediately. Inventory requirements for 125 SKUs (presently occupying about 3,590 pallet positions in average inventory) will be

**FIGURE 72**

**Design Parameters for Cross Docked Items**

| PLANNING YEAR | Year 1 | Year 3 |
|---|---|---|
| **Projected Products to Cross Dock** (SKUs) | 125 | 690 |
| SKUs cross docked in pallet loads | 25 | 82 |
| SKUs cross docked in full cartons | 100 | 608 |
| **Projected Number of Suppliers** | 24 | 36 |
| **Projected Average Inventory Eliminated** (pallet positions) | 3,590 | 10,350 |
| **Projected Value of Inventory Eliminated** | $1,448,051 | $3,131,058 |
| **Projected No. of Stores** (approx 15% growth) | 83 | 100 |
| **Stores Serviced per Day** | 17 | 20 |
| **Projected Average Throughput** (pallets/day) | 78 | 155 |
| Avg full pallets cross docked (pallets/day) | 16 | 33 |
| Avg SKUs Per Day in full pallets (lines/day) | 8 | 25 |
| Avg full cartons cross docked (cartons/day) | 1,479 | 2,944 |
| Avg SKUs per day in full cartons (lines/day) | 21 | 113 |
| Avg pallets of full cartons cross docked (pallets/day) | 62 | 123 |
| **Average Cross Docked Pallets Per Store** | 4.6 | 7.8 |

reduced or eliminated. By Year 3, cross dock volume increases to 690 SKUs (10,350 pallet positions) from 36 suppliers with an average throughput of about 155 pallets cross docked daily. The Year 3 volume is the design capacity for the system.

### Step 2: Develop alternative operational requirements

Based on these projected design parameters, S&O developed alternatives for order procedures and cross docking operations.

**Order Procedure Alternatives**

Saw & Order is contemplating two different methods of placing orders for the cross docked items.

**Order Alternative 1.** The simplest procedure to implement provides store managers with a list of products to be cross docked, the dates on which the material will be received, and the deadlines for ordering each item. The store manager will also be provided with other relevant information, such as minimum order quantities of each item, or whether promotions for any items are scheduled.

Each manager will be required to order the quantity he wants to receive on his store's delivery date *by the deadline*. The deadline will be determined by the lead time of the supplier to the warehouse, transit time from the warehouse to the store, the scheduled day of the week for store delivery, and the time needed to accumulate and consolidate the order information from the stores. Purchasing will tally the orders for products in the quantities specified in a consolidated list of store manager orders. When the products are received, the orders will be released to the warehouse for sorting as described in the proposed facility operation.

Under this ordering scenario, a policy needs to be developed to handle the remainder quantities for items that are received from the supplier in pallet quantities, but ordered from the store managers in carton quantities. Those items may either be distributed to the stores or held in the warehouse and deducted from the next consolidated order for that item. When deliveries are short, the quantities shipped to the store must be reduced by a formula acceptable to management. Some options include reducing the quantity allotted to each store by an equal percentage, giving higher volume stores a larger percentage of their requests, basing the decision on demographic considerations, or any other method of allocation that seems appropriate.

**Order Alternative 2.** The second method involves a more complicated ordering process, but is invisible to the store manager. In this concept, the central ordering department is responsible for producing order forecasts for the products to cross dock. These order forecasts will be based on the knowledge of the marketplace, history of the movement of the product, and knowledge of future promotions. The suppliers must be part of a cooperative effort to make this work. Deliveries are on a weekly schedule, and quantities may vary from week-to-week depending on the changes in the marketplace, and promotional plans.

At the store level, the manager will continue to place orders as he does now, with the only difference being timing. When all store orders are consolidated, any difference between the quantity expected, and the total quantity ordered by the stores will be treated as a warehouse transaction. If the quantity ordered by the stores is larger than the quantity to be received at S&O, the difference will be made up from a small inventory that is carried in the warehouse to handle week-to-week variations. Because the products selected for cross docking have been limited to those that have reasonably predictable sales patterns, this difference should be at a minimum. The average quantity stored in the warehouse should be no larger than one-half of one week's average sales. As the program progresses, this quantity may be adjusted on an item-by-item basis.

**Order Alternative 1 versus Alternative 2.** Each of these methods has advantages and disadvantages. The first alternative will eliminate inventory in the warehouse for items that are cross docked. However, it will place responsibility for sales forecasting and inventory management on the store manager. This may lead to extended stock outs on some items, and high inventories on others, resulting in increased store-to-store transactions.

The second method eliminates the store manager as the decision maker. However, it implies that a small inventory of cross docked items will be maintained in the warehouse, and it also requires someone to monitor changes in sales patterns and warehouse inventories, so that the proper balance of incoming shipments and warehouse inventory can be maintained. In order for the program to be effective, forecasting the stores' needs must be accurate. If the needs for particular products are overestimated, they will have to be warehoused, negating the benefits of a cross docking operation.

For either of the alternatives to work, the person responsible for dealing with the suppliers must maintain a close relationship with them. S&O and the supplier must be a team. The supplier must be aware of upcoming promotions as far in advance as possible so that the desired quantity of a product will be available. In turn, the supplier must tell S&O about any delays or shortages. Delays must be communicated to the warehouse and to MIS in addition to the stores so that necessary changes in the release of orders can be made. When the delay is less than one week, only the stores for which delivery will be delayed need to be notified. Shortages are handled by reducing the quantities shipped to stores in accordance with a formula acceptable to store management. Large over-shipments should be returned immediately to the supplier.

Alternative 2 requires weekly deliveries from the suppliers. Fewer deliveries would mean that more product needs to be warehoused between deliveries.

**Cross Docking Operations Alternatives**
Based on the design parameters, it was determined that SKUs that moved in *full pallet quantities* and were pre-labeled by the supplier would be transferred from receiving to shipping staging using existing counterbalanced lift trucks. A receiving person would scan the bar code on an incoming pallet and on-board radio frequency terminals would direct him to the appropriate staging lane or dock door. The rest of the product would have to be manually sorted at the carton level. For cross docking these *full cartons*, the following alternatives were generated.

**Operations Alternative 1: Sort-to-Pallet.** The study team decided that sort-to-pallet was one option for processing carton quantities. To prevent congestion and confusion in the regular staging area, space would be provided in the cross dock area for pallets to be staged while waiting to be sorted. As a SKU is unloaded, and the receiving function completed, a distribution list is created for that SKU, indicating the quantity to be distributed to each store. This list should be generated by product regardless of how many different products are received on a trailer, because sorting will be done on a product-by-product basis. In addition, the list should be printed in the same sequence that the lanes are assigned to stores in the sortation aisles.

Using an electric pallet jack, a warehouse person will take one pallet at a time, and travel along the cross dock aisles, placing the appropriate number of cartons on the empty pallet in front of the rack lane for each store. This process continues for each cross docked item received. When the pallet on the floor is filled, it is wrapped, labeled by store, and lifted into the drive-in racks to the next available position, and a new empty pallet is placed on the floor. Including the pallet position in front of the rack, each lane will hold 9 pallet positions. Although design parameters indicate about 7.8 pallets are to be delivered to each store weekly, an average of 2 pallets per store do not need to be sorted at the carton level, but are transferred from receiving to shipping in full pallet quantities. Stores with higher volumes may be assigned two lanes.

**Operations Alternative 2: Pick-From-Conveyor.** The second alternative for sorting cross docked products is slightly more automated and uses conveyors. Unit loads that need to be sorted by

# CHAPTER 6 A Case Study on Cross Docking

stores at the carton level need not be removed from a truck's trailer. Instead of a distribution list, actual store labels are printed out for the cartons on the pallet. Labels may be color coded for a specific set of stores with store numbers printed in large, easy-to-read text for quick identification. The receiver/unloader will attach a label to each carton and place it on an extendable conveyor for transport to a manual sortation area, which is made up of into rack lanes that flank the incoming conveyor. Each lane is permanently assigned to a single store and is equipped with pallet flow racks at the bottom level and push back racks at the top two levels (Figure 73). In front of the lanes, a warehouse person, assigned to a zone consisting of multiple adjacent store lanes, takes the color-coded carton for the assigned zone off the conveyor and places it on an empty pallet at the bottom level of the specific store rack lane. The process will continue for all the cartons received for sortation. When a pallet is filled to capacity, the sorter pushes it to the back position on the flow rack and empty pallet is placed in the front. The full pallet is wrapped, labeled by store, and lifted into the next available position on the upper level push back racks. A recirculating conveyor allows a carton that was not picked on the initial pass to be conveyed back to the sorter for another chance—the same concept used in baggage claim systems in airports.

**FIGURE 73**

**Conceptual View of a Store Lane for Case Study**

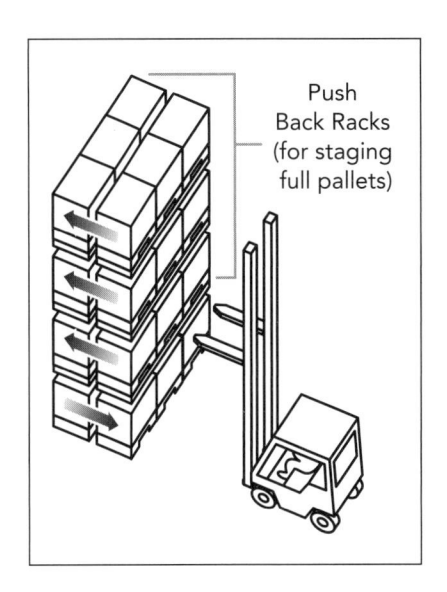

**FIGURE 74**

**Existing Layout for Case Study**

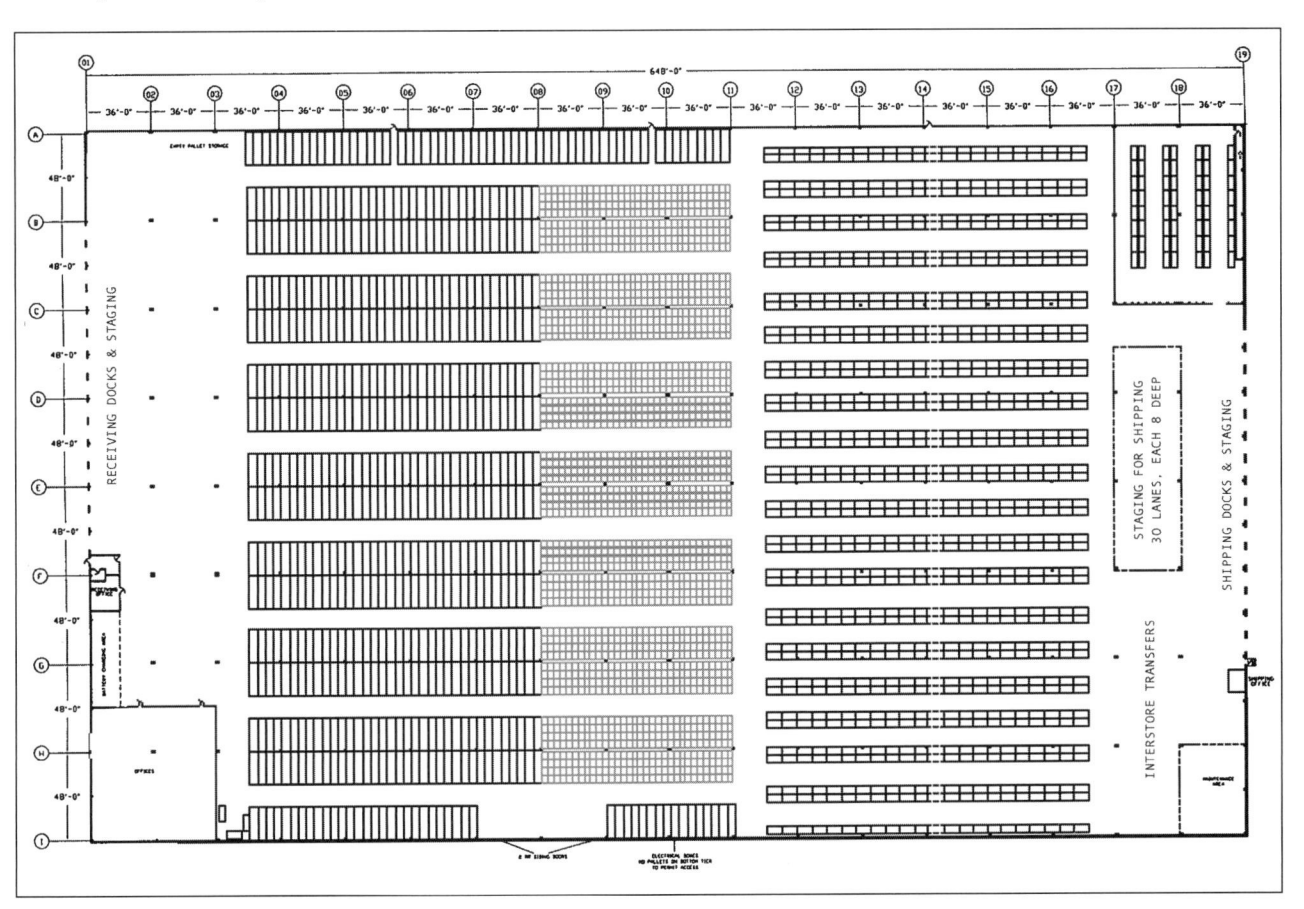

### Step 3: Lay out the different alternatives

Computer-aided design (CAD) software was used to lay out the different cross docking alternatives within the existing facility. Figure 74 illustrates the existing facility where receiving functions are at one end of the building and shipping functions are at the other end. This flow may be ideal for traditional store-then-ship operations, but it is clearly a liability for cross docking's fast-paced nature. S&O agreed that the *cross dock receiving function* for both alternatives should be situated at the shipping end of the building. By doing so, the long run across the entire length of the building to move incoming pallets from receiving to shipping was eliminated. Two dock doors, originally for shipping, will be designated as receiving doors for cross docked items. For both alternatives, pallet racks were removed. With cross docking, inventory is reduced and the removal of the racks was anticipated. This is not expected to encroach on a lower storage capacity requirement for the facility. Figure 75 illustrates the layout for Alternative 1, the sort-to-pallet approach. Lanes for approximately 100 stores are provided. A cross docking receiving staging area is necessary for pallets to be staged while awaiting

**FIGURE 75**

**Alternative 1 Layout for Case Study**

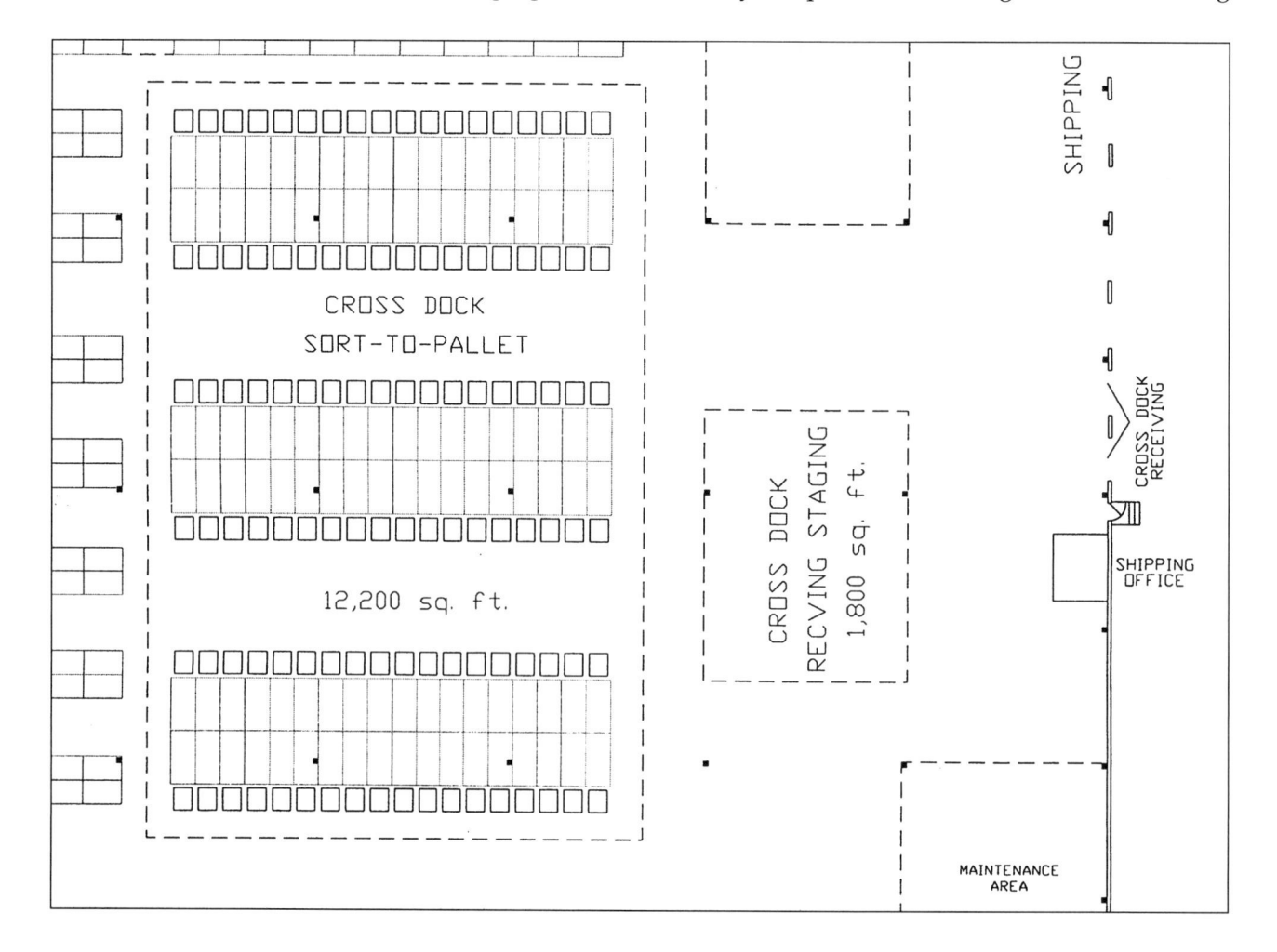

CHAPTER 6   *A Case Study on Cross Docking*

sorting. Figure 76 illustrates the layout for Alternative 2, the pick-from-conveyor scenario. To avoid long runs by the recirculating conveyor, sortation was divided into two main sections with lanes provided for 50 stores for each section for a total of 100 lanes. A warehouse person at the end of the first section will be needed to manually divert cartons meant for the other set of stores. A set-up area is included to stage pallets of items that may have been in temporary storage, but needed to fill orders to stores slated for delivery that day. Store labels are printed out in a labeling area adjacent to this staging area. A warehouse person attaches store labels to the cartons and places them on the sortation conveyor.

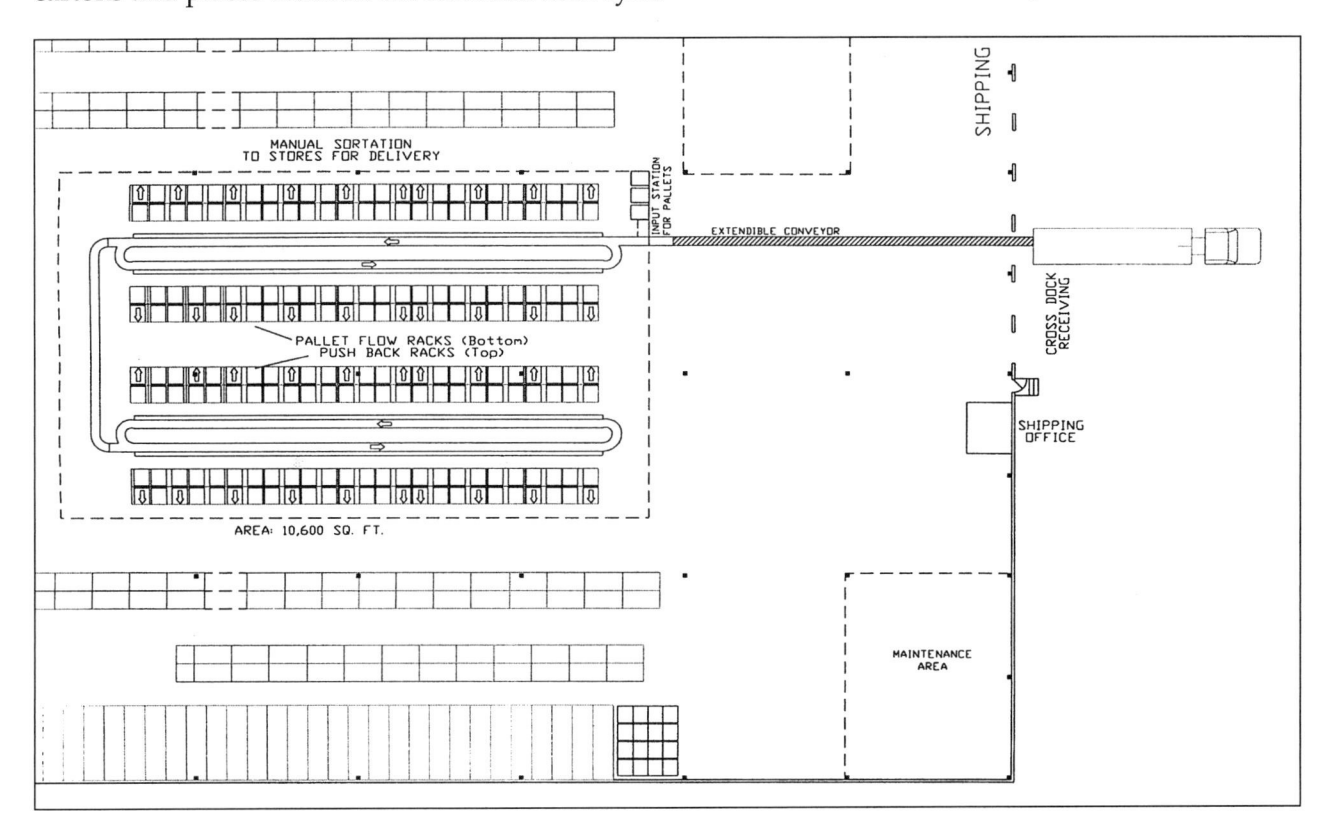

**FIGURE 76**

**Alternative 2 Layout for Case Study**

### Step 4: Develop labor and equipment requirements

Based on the design parameters, labor and equipment requirements were developed for each alternative (Figure 77). When comparing the labor requirements between the two operations alternatives, S&O only considered the incremental difference of one alternative versus the other to simplify the costing analysis. The goal was to determine which operation would better suit S&O's needs and provide a greater return on their investment. For example, the full pallet moves for both operations alternatives were the same, thus the labor requirements for that aspect of the operation was not taken into consideration at that time. For Alternative 1, eight people on electric pallet jacks will be needed to distribute cartons to the stores. Two people will be needed to move pallets from the truck to the staging area for the sorters, while an additional lift truck driver will be needed to move completed pallets from the front of a store lane to the drive-in racks for staging. For Alternative 2, with the aid of conveyors, people sorting to stores is reduced to four, with two additional persons labeling and unloading cartons at the dock door as the product arrives. At least one lift truck driver will be needed to move completed pallets from the bottom level pallet flow racks to the upper level push back racks. For the carton-level cross dock operation, 11 people will be needed for Alternative 1 versus 7 people for Alternative 2. Equipment requirements (number of pallet positions, total electric pallet jacks, total length of conveyors, etc.) are determined from the layout and from the labor requirements. S&O will re-use two electric pallet jacks for Alternative 1. The remainder of the equipment will be purchased new.

**FIGURE 77**

Labor Requirements for the Cross Dock of Full Cartons Only

|  | Standard Minutes Per Move[1] | No. of Moves Per Day | Total Minutes | Total Personnel Required[2] |
|---|---|---|---|---|
| **Alternative 1** | | | | |
| Sort-to-Pallet Personnel (Using Pallet Jack) | | | | |
|     Label carton and arrange on pallet | 0.38 | 2,944 | 1,119 | |
|     Move to next store position | 0.86 | 2,263 | 1,946 | |
|     **Total** | | | 3,065 | 8 |
| Pallet Moves (Using CB Lift Truck) | | | | |
|     Move from receiving dock to receiving staging | 3.51 | 123 | 430 | 2 |
|     Lift from sort to rack staging | 2.34 | 123 | 287 | 1 |
| **Total Required for Cross Dock Operation 1** | | | | **11** |
| **Alternative 2** | | | | |
| Unloading Personnel | | | | |
|     Label carton and place on conveyor | 0.23 | 2,944 | 677 | 2 |
| Pick-From-Conveyor Personnel | | | | |
|     Obtain carton and arrange on pallet | 0.25 | 2,944 | 736 | |
|     Walk to store lanes in zone (approx 20 ft) | 0.21 | 2,263 | 476 | |
|     Manual Divert | 0.23 | 1,472 | 339 | |
|     **Total** | | | 1,551 | 4 |
| Pallet Moves (Using CB Lift Truck) | | | | |
|     Lift from flow rack to rack staging | 2.00 | 123 | 245 | 1 |
| **Total Required for Cross Dock Operation 2** | | | | **7** |

1 From time studies and time output from WMS transactions, includes information tracking, PF&D and other minor movements.
2 Based on a 7-hour workday and rounded up to the nearest integer.

## Step 5: List information systems requirements

The information systems support required with a cross dock program is essentially the same for both alternatives. MIS will be required to keep information on quantities expected each week. It must maintain information for each item, the store, by the week, and the quantity to be shipped. By doing this for each store, there will be a list of records with two elements, the week for which the material is requested, and the quantity needed. The store file must also contain the day of the week when the store is serviced.

Another file of open purchase orders with delivery dates and quantities must be maintained, and must be updated whenever the supplier informs S&O of its inability to meet the requested quantity or delivery date.

On the day the material is scheduled to arrive, a distribution list that provides the store number and the quantity to be placed on that store's pallet must be printed for each item. To produce the list, the computer searches the list of open requests to locate requests with the oldest need date. Within this group, the stores with the closest replenishment day will be serviced first, and within the same replenishment day, the sort will be by the store number. Once this sort is completed, the computer will set the expected arrival quantity to the available quantity and:

1. Examine the next store request record.
2. Add the quantity on the record to the quantity to be shipped to the store number on the same record.
3. Reduce the available quantity by the quantity of the record.
4. Mark the store request record as fulfilled.
5. See if the available quantity is greater than zero.
6. If the available quantity is greater than zero, repeat the cycle.
7. If the available quantity is less than the amount ordered for the stores because of an under shipment, reallocate the merchandise to the stores, based on the agreed-upon formula for reducing the quantities. (Under Alternative 2, the shortage should be made up first from the smaller warehouse stock, and then this procedure is followed if necessary.)
8. When the available quantity reaches zero, print the distribution list.

*In this way, the oldest store needs will be serviced first while ensuring that the material is moved through the facility as quickly as possible.*

There are a fewer minor differences in the information systems support for the operations alternatives.

- Alternative 1: Requires creating of a distribution list to be used by sort-to-pallet personnel.
- Alternative 2: In lieu of a distribution list, requires color-coded store labels with large, easily-read store numbers. In addition,

some method of assigning and balancing work in the sortation zones is needed.

## 6.3 Economic analysis of alternatives

The next part of the process evaluates and compares the different scenarios quantitatively and qualitatively.[5]

### Step 1: Assemble mutually-exclusive alternatives

Because the operations alternatives have a greater impact on capital investment, these alternatives are the ones that S&O takes into consideration in the economic analysis.

### Step 2: Calculate the capital costs of each alternative

Figure 78 lists the equipment and capital investment required for each alternative. The information systems upgrade is an estimated amount provided by the company's WMS provider for both scenarios. Two existing electric pallet jacks will be re-used for Alternative 1, thus reducing its total capital costs to $331,120. Capital costs for Alternative 2 total $533,200.

### Step 3: Calculate the operating costs of each alternative

Because S&O wants to determine which alternative is more feasible, the focus is not on the alternatives' common costs, *but the difference*

**FIGURE 78**

**Capital Costs of Equipment and Information Systems**

|  |  |  | ALTERNATIVE 1 |  | ALTERNATIVE 2 |  |
|---|---|---|---|---|---|---|
| COST CATEGORY | Unit | Unit Cost | Units Provided | Total Cost | Units Provided | Total Cost |
| **Equipment** |  |  |  |  |  |  |
| Racks: |  |  |  |  |  |  |
| Drive-in Rack | PAL POS | $ 80 | 864 | $ 69,120 | N/A | $ – |
| Pallet Flow Rack | PAL POS | $ 250 | N/A | $ – | 200 | $ 50,000 |
| Push Back Rack | PAL POS | $ 170 | N/A | $ – | 600 | $ 102,000 |
| Trucks: |  |  |  |  |  |  |
| Electric Pallet Jack | EACH | $ 9,500 | 8 | $ 76,000 | N/A | $ – |
| Existing Pallet Jack (to be re-used) |  |  | 2 | $ (19,000) | N/A | $ – |
| CB Lift Truck | EACH | $25,000 | 3 | $ 75,000 | 1 | $ 25,000 |
| Conveyor: |  |  |  |  |  |  |
| Powered Conveyor | FOOT | $ 300 | N/A | $ – | 514 | $ 154,200 |
| Extendible Conveyor | EACH | $27,000 | N/A | $ – | 1 | $ 27,000 |
| **Facility Upgrades** |  |  |  |  |  |  |
| Rack Removal |  |  |  | $ 30,000 |  | $ 20,000 |
| Other Utility Upgrades |  |  |  | $ – |  | $ 5,000 |
| **Information Systems** |  |  |  |  |  |  |
| Changes/Upgrades |  |  |  | $ 100,000 |  | $ 150,000 |
| **Total Capital Costs** |  |  |  | **$ 331,120** |  | **$ 533,200** |

1 All costs include costs for delivery, installation, and start-up.

CHAPTER 6  *A Case Study on Cross Docking*   159

*in their incremental costs.* The main difference in operating costs lies in labor costs. Four fewer people are required for Alternative 2 at a savings of $35,000 (including benefits and insurance) per person. Figure 79 summarizes the annual operating costs for labor, space, equipment, and information systems.

**FIGURE 79**

**Annual Operating Costs**

|  |  |  | ALTERNATIVE 1 | | ALTERNATIVE 2 | |
|---|---|---|---|---|---|---|
| COST CATEGORY | Unit | Annual Cost Per Unit | Units Provided | Annual Cost | Units Provided | Total Cost |
| **Labor**[1] | PERSON | $ 35,000 | 11 | $ 385,000 | 7 | $ 245,000 |
| **Storage**[2] | SQ FOOT | $ 8 | 14,000 | $ 112,000 | 10,600 | $ 84,800 |
| **Equipment**[3] | | | | | | |
| Racks: | | | | | | |
|    Drive-in Rack | PAL POS | $ 0.2 | 648 | $ 130 | N/A | $ - |
|    Pallet Flow Rack | PAL POS | $ 0.6 | N/A | $ - | 200 | $ 120 |
|    Push Back Rack | PAL POS | $ 0.4 | N/A | $ - | 400 | $ 160 |
| Trucks: | | | | | | |
|    Electric Pallet Jack | EACH | $ 443 | 8 | $ 3,544 | N/A | $ - |
|    CB Lift Truck | EACH | $ 801 | 3 | $ 2,403 | 1 | $ 801 |
| Conveyor: | | | | | | |
|    Powered Conveyor | FOOT | $ 6 | N/A | $ - | 514 | $ 3,084 |
|    Extendible Conveyor | EACH | $ 540 | N/A | $ - | 1 | $ 540 |
| **Information Systems**[4] | | | | $ 30,000 | | $ 32,500 |
| **Annual Operating Costs** | | | | **$ 533,077** | | **$ 367,005** |

1 Includes salaries, benefits, and insurance.
2 Expenses include annual costs of ownership and operation.
3 Expenses include annual costs of operation (utility consumption, maintenance and repairs)
4 Includes half a person's salary to monitor movement of cross docked items.

## Step 4: Choose a method for evaluating and comparing investment alternatives

Two methods of economic analysis were used. The **payback period** method enables management to determine the liquidity of investing in cross docking based on the savings gained versus the present system. However, management recognized that the payback period method ignores the long-term effects of the investment. To determine which alternative would be more feasible for the long term, the equivalent uniform annual cost (EUAC) approach was used.[6] Figure 80 illustrates the results of both methods. The company's finance department provided salvage values and interest rates to use in the EUAC analysis. A 10-year economic planning life was used because this was the length of time that S&O wanted to stay in the facility. The EUAC method shows that Alternative 2 has 25 percent lower annualized costs than Alternative 1. A simple comparison of costs was performed against the existing store-and-ship method.

**FIGURE 80**

**Economic Analysis of Saw & Order**

It shows a 51 percent reduction in annual costs for Alternative 2 versus existing methods. The payback periods for both alternatives are the same at a year.

| METHOD | Existing | ALTERNATIVE 1<br>Sort-to-Pallet | ALTERNATIVE 2<br>Pick-From-Conveyor |
|---|---|---|---|
| **Payback Period Method** | | | |
| 1. Total Capital Investment (Figure 78) | $ — | $ 331,120 | $ 533,200 |
| 2. Difference from Base | base | $ 331,120 | $ 533,200 |
| 3. Annual Operating Costs (Figure 79) | $ 878,256 | $ 533,077 | $ 367,005 |
| 4. Savings from Base | base | $(345,179) | $(511,251) |
| 5. Payback Period in Years (line 4 ÷ line 2) | | **1.0** | **1.0** |
| **Annual Cash Flow Method** | | | |
| 1. Total Capital Investment (Figure 78) | | $ 331,120 | $ 533,200 |
| 2. Planned Economic Life | | 10 | 10 |
| 3. Estimated Salvage Value at End of Economic Life | | $ 198,672 | $ 373,240 |
| 4. MARR | | 10% | 10% |
| 5. Capital recovery factor (i = 10%) | | 0.16275 | 0.16275 |
| 6. Sinking fund factor (i = 10%) | | 0.06275 | 0.06275 |
| 7. Equiv. Annual Cost of Capital Recovery<br>([line 1 × line 5] − [line 3 × line 6]) | | $ 41,423 | $ 63,357 |
| 8. Annual Operating Cost (Figure 79) | $ 878,256 | $ 533,077 | $ 367,005 |
| 9. **Total Equivalent Annual Cost (line 7 + line 8)** | **$ 878,256** | **$ 574,500** | **$ 430,362** |

*25% reduction in annual costs for Alt 2 vs. Alt 1*

*51% reduction in annual costs for Alt 2 vs. Existing*

## Step 5: Perform a qualitative analysis

A qualitative analysis further emphasized the advantages of Alternative 2 over Alternative 1. Although Alternative 1 would be easier to implement with only drive-in racks needed for installation, Alternative 2 requires fewer people to manage and is more flexible for changing business conditions and volume fluctuations. In Alternative 2, to make room for more pallets per store in a lane, store sortation need not be stopped when unloading completed pallets from the push back racks. For Alternative 1, sorting to a store would have to be suspended when emptying a store lane.

## Step 6: Specify the selected alternative

Based on a quantitative and qualitative analysis, S&O decided that Alternative 2, the pick-from-conveyor scenario, would be adopted for cross docking at the carton level. At the pallet level, cross docking would be the transfer of pallets from receiving or receiving staging directly to shipping staging or shipping.

CHAPTER 6   A Case Study on Cross Docking    161

S&O also decided to implement Order Alternative 2 in which the cross dock operation is invisible to the store manager and a central ordering department is responsible for forecasting and inventory management. Although the need for a higher inventory is implied with Order Alternative 2, it does not place the responsibility of sales forecasting and inventory management on the store manager, which avoids placing unnecessary inventory or "forecasting mistakes" on a store's more expensive retail space and reduces store-to-store transactions.

## 6.4   Implementation

A detailed implementation plan was created and a pilot program of three suppliers and a handful of SKUs was implemented immediately with full pallets and no major difficulty.

The site was prepared for the manual carton sortation system. Pallet racks were removed and conveyors were purchased and installed. The case-level portion of the program was activated within six months.

A couple of stores reported stock-outs of one or two cross docked items, but this was determined to be due to flaws in forecasting the store's needs and not a problem with the cross dock operation itself. In the future, the company hopes to use Point-of-Sale (POS) data and computer-aided ordering (CAO) to improve their ordering and forecasting functions.

There were some work balancing issues that had to be programmed into S&O's WMS specifically for the manual sortation system. Information systems testing took slightly longer than anticipated, but did not delay the overall project.

The order and inventory management function was accustomed to making a distribution after the actual product was received. With cross docking, it had to have a different mindset and get used to distributing against expected receipts for items slated for cross docking. This problem was quickly remedied in training sessions that emphasized the store destination of a cross docked item must be determined before the actual product is received.

Other than these challenges, S&O did not encounter any major implementation difficulties.

## 6.5   One year of operation

After a year of operations, S&O received the benefits of cross docking on schedule and on budget. As expected, its decision to cross dock did not affect the overall quantity of product that went through the warehouse because it did not expect any increased sales due to cross docking. S&O felt that it accomplished the following:

- *Reduced handling.* Under the old store-and-ship system, items underwent a five-step process from receiving to shipping. Merchandise was received, put away in the storage area, moved to the appropriate picking area, picked onto pallets, and shipped to the stores. With the cross dock system, two steps were eliminated for those same items. Items are now received, put away directly onto store pallets, and shipped. The annual operating cost savings from cross docking were even better than expected—over $600,000.

- *Decreased inventory investment.* For the cross docked items, the average inventory in the past was 6.9 weeks. With cross docking, there is never more than 1.5 weeks of inventory.

- *Reduced storage space requirements.* By the end of Year 1, over 3,500 pallet positions were vacated from inventory. The cross dock operation used the space for about 1,500 pallet positions, which left a net gain of 2,000 available pallet positions. In addition, the number of pallets in the carton-level cross dock averaged under four pallets per store—or only 47 percent of the designed capacity of eight pallets. With the projected increase in cross docking by Year 3, store lanes will still only be at 76 percent of capacity. If any additional products in the warehouse are found to be cross dockable, additional warehouse storage positions would be vacated while extra dock space would not be needed. Even greater space savings would result.

- *Improved supplier and S&O relations.* Suppliers that participated successfully in the program were able to increase their product offerings to S&O. In some cases, prices for their goods may be slightly higher than their competitors, but because they were part of the program, S&O preferred to deal with them and to include their products in its cross docking program. One supplier, however, had to withdraw from the program because of its inability to provide consistently correct and accurate shipments in a timely manner.

## 6.6 The future for S&O

- *Expansion of the program.* Because of its success with the program, S&O is considering cross docking some items that are shipped directly to its stores. These were not included in the initial evaluation because management had decided to concentrate only on those items moving through the warehouse. The benefits of cross docking these direct-shipped items include the possibility of lower prices and freight rates from consolidated orders and greater control from a centralized ordering system.

It will, however, impact on dock capacity, as a higher number of receipts and shipments will need to be processed.

- *Negotiation with suppliers.* The current system does not rely on the supplier's ability to pre-allocate and consolidate product by stores. Thus most of the cross docking involves the need to sort incoming cartons by stores at S&O's facility. In the future, S&O plans to negotiate with its suppliers to increase the number of pre-allocated supplier-consolidated pallets that can be cross docked through the facility. The full pallet movement, characteristic of supplier consolidation, will further decrease handling requirements in the warehouse because sortation for those items will be eliminated.

- *Adoption of best practices pull distribution.* Current S&O cross docking efforts depend on forecasts—not on actual demand data. S&O is looking to synchronize its entire supply chain activities with actual consumer demand by using Point-of-Sale information and computer-aided ordering processes to create a dynamic cross docking environment. Actual sales will update a perpetual inventory of each item for each store. When in-store inventory of an item falls below a carefully established reorder point (based on store demographics, projected weekly promotional or merchandising activities, and manufacturer's lead times), replenishment orders will be automatically generated and aggregate quantities per item will be calculated for a specific set of stores and transmitted to suppliers. Suppliers will ship aggregate quantities of SKUs for a specific set of stores to S&O's facility so that it can arrive within a day or two of when those stores are to be replenished.

- *Expansion and upgrade to a new facility.* After a single year of operation, S&O saw sales rise three times more than anticipated. They are planning to expand and upgrade to a new facility in four years instead of eight. They expect this new facility to have an automatic sortation system to facilitate sorting cartons to their growing number of stores.

## Notes

[1] Not a real company name.
[2] Product and supplier selection used the approach presented in Chapter 2, Sections 2.3 and 2.4.
[3] For a comprehensive discussion on assessing current operations and facilities and their propensity for cross docking, refer to Chapter 2, Section 2.5.
[4] Based on a procedure described in Chapter 3, Section 3.3.
[5] Based on the procedure described in Chapter 3, Section 3.4.
[6] The EUAC and other methods of economic analysis are described in greater detail in Chapter 3, Section 3.4.

# APPENDIX A

# Organizations & Information Sources for Cross Docking

Listed on the following pages are some trade and professional organizations in logistics, warehousing, and physical distribution.[1] These organizations can provide prospective cross docking practitioners with information regarding equipment, information systems, and other relevant material pertaining to cross docking.

For a comprehensive list of software programs with cross docking capability, see *Logistics Software* by Andersen Consulting and prepared for the Council of Logistics Management (CLM).

For more information of modeling distribution networks and site selection, see:

- Napolitano, Maida, *Using Modeling to Solve Warehousing Problems*, Warehousing Education and Research Council, Chicago, 1998. (Section on network modeling and site location includes a comprehensive list of references.)

- Robeson, James, and Copacino, William (eds.), *The Logistics Handbook*, The Free Press, New York, 1994.

For more information on GANTT charts and project scheduling:

- Gordon, Michele, et al., *Microsoft Project 98 Step-by-Step with CDRom*, Microsoft Press, 1997.

- Lewis, James P., *Project Planning, Scheduling and Control: A Hands-On Guide to Bringing Projects In On Time and On Budget*, The McGraw Hill Companies, New York, 1995.

## Notes

1 Adapted from a list of organizations compiled by the Council of Logistics Management. For a more complete list, contact the Council of Logistics Management, Oak Brook, IL

| Organization Name | Purpose/Objective | Address | Phone No. | Website |
|---|---|---|---|---|
| American Production & Inventory Control Society (APICS) | Develop professional efficiency in resource management | 5301 Shawnee Rd, Alexandria, VA 22312 | 703-354-8851 | http://www.apics.org/ |
| American Productivity & Quality Center (APQC) | Increase productivity and quality, emphasizing total quality management | 123 North Post Oak Lane, Houston, TX 77024 | 713-681-4020 | http://www.apqc.org/ |
| American Society of Transportation & Logistics, Inc. (AST&L) | Establish, promote, and maintain standards of knowledge and professional training | 229 Peachtree St., Atlanta, GA 30303 | 404-524-3555 | http://www.astl.org/ |
| American Trucking Associations, Inc. (ATA) | Serve united interests of the trucking industry | 2200 Mill Rd., Alexandria, VA 22314 | 703-838-1700 | http://www.trucking.org/ |
| Association of American Railroads | To better enable the railroad to operate as a national system | 50 F Street NW., Washington DC 20001 | 202-639-2100 | http://www.aar.org/ |
| Association of Professional Material Handling Consultants (APMHC) | Develop standards of performance | 8720 Red Oak Blvd., Charlotte, NC 28217 | 704-676-1184 | |
| Canadian Association of Supply Chain & Logistics Management | Improve logistics and distribution management skills | 590 Alden Rd., Ste. 211, Markham, Ontario L3R 82N | 905-513-7300 | http://www.infochain.org |
| Canadian Industrial Transportation Association (CITA) | Promote, conserve, protect commercial transportation interests | 75 Albert St., Ste. 1002, Ottawa, ON K1P 5E7 | 613-235-2482 | http://www.citl.ca |
| Council of Logistics Management (CLM) | Provide leadership in defining and understanding the logistics process | 2805 Butterfield Rd., Suite 200, Oak Brook, IL 60523 | 630-574-0985 | http://www.clm1.org/ |
| Food Distributors International | Improve efficiency and increase profitability of wholesale grocery industry | 201 Park Washington Court, Falls Church, VA 22046 | 703-532-9400 | http://www.fdi.org |
| Grocery Manufacturers of America, Inc. (GMA) | Solve problems in the grocery industry | 1010 Wisconsin Ave. NW, Fl 9, Washington, DC 20007 | 202-337-9400 | http://www.gmabrands.com/ |
| Institute of Industrial Engineers (IIE) | Assist research in the field of industrial engineering | 25 Technology Park Atlanta, Norcross, GA 30092 | 800-494-0460 | http://www.iienet.org/ |
| Institute of Management Consultants, Inc. (IMC) | Provide professional certification to assure standards and qualification | 2025 M Street NW, Ste. 800, Washington, DC 20036 | 202-367-1134 | http://www.imcusa.org/imc.html |
| Institute of Packaging Professionals (IoPP) | Assist professional development of packaging and handling professionals | 481 Carlisle Dr., Herndon, VA 20170 | 703-318-8970 | http://www.iopp.org/ |
| Intermodal Association of North America | Promote intermodal transportation | 7501 Greenway Ctr. Dr., Ste. 720, Greenbelt, MD 20770 | 301-982-3400 | http://www.intermodal.org/ |

# APPENDIX A  Organizations and Information Sources for Cross Docking

| Organization Name | Purpose/Objective | Address | Phone No. | Website |
|---|---|---|---|---|
| International Association of Refrigerated Warehouses | Advance interest in refrigerated warehouse business | 7315 Wisconsin Ave., Suite 1200N, Bethesda, MD 20814 | 301-652-5674 | http://www.iarw.org/ |
| International Customer Service Association (ICSA) | Develop understanding of total quality service process | 401 N. Michigan Ave., Chicago IL 60611 | 312-321-6800 | http://www.icsa.org |
| International Mass Retailing Association (IMRA) | Represent nation's discount and variety general merchandise retail industry | 1700 N Moore St., Ste. 2250, Arlington, VA 22209 | 703-841-2300 | http://www.imra.org |
| International Society for Inventory Research | Provide framework for dissemination of research results | Veres Palne u. 36, Budapest, Hungary H-1053 | 36-1-317-2959 | |
| International Warehouse Logistics Association (IWLA) | Conduct research to improve warehouse efficiency | 1300 Higgins Rd., Park Ridge, IL 60068 | 847-292-1891 | http://www.warehouselogistics.org |
| Material Handling Equipment Distributors Association | Educate in methods and practices for efficient materials handling | 201 Route 45, Vernon Hills, IL 60061 | 847-680-3500 | http://www.mheda.org/ |
| The Material Handling Institute, Inc. (MHI) | Provide materials handling information | 8720 Red Oak Blvd., Charlotte, NC 28217 | 704-676-1190 | http://www.mhia.org/mhi/ |
| Materials Handling & Management Society | Promote public recognition of material handling and material management | 8720 Red Oak Blvd., Suite 210, Charlotte, NC 28217 | 704-676-1199 | |
| The National Association of Manufacturers (NAM) | Promote America's economic health and productivity in the manufacturing sector | 1331 Pennsylvania Ave. NW, Washington, DC 20004 | 202-637-3000 | http://www.nam.org/ |
| The National Association of Purchasing Management (NAPM) | Provide leadership in purchasing and materials management | PO Box 22160, Tempe, AZ 85285 | 800-888-6276 | http://www.napm.org/ |
| The National Private Truck Council (NPTC) | Represent interests of private trucking industry before judiciary branch of government | 66 Canal Center Plaza, Suite 600, Alexandria, VA 22314 | 703-683-1300 | http://www.nptc.org/ |
| The National Small Shipments Traffic Conference (NASSTRAC) | Identify innovative ways to save money during distribution of goods | 499 S. Capitol St. SW, Ste. 604, Washington, DC 20003 | 202-484-9188 | http://www.nasstrac.org |
| Society of Logistics Engineers (SOLE) | Enhance art and science of logistics technology | 8100 Professional Place, Suite 211, Hyattsville, MD 20785 | 301-459-8446 | http://www.sole.org/ |
| Transportation Research Board (TRB) | Advance knowledge concerning nature and performance of transportation systems | 2101 Constitution Ave., NW, Washington, DC 20418 | 202-334-2934 | http://www.nationalacademies.org |
| Transportation Research Forum | Provide an impartial meeting ground for shippers to exchange information | 1 Farragut Square, Ste. 500, Washington, DC 20006 | 202-879-4701 | http://www.utexas.edu/depts/ctr/trf/ |

| Organization Name | Purpose/Objective | Address | Phone No. | Website |
|---|---|---|---|---|
| Uniform Code Council, Inc. | Promote standards and service that support product identification and electronic data interchange | 7887 Washington Village Dr., Ste. 300, Dayton, OH 45459 | 937-435-3870 | http://www.uc-council.org/ |
| US Chamber of Commerce | Enhance human progress through better system efficiency | 1615 H Street NW, Washington, DC 20062 | 202-659-6000 | |
| Warehousing Education and Research Council (WERC) | Provide education and to conduct research concerning the warehousing process | 1100 Jorie Boulevard Suite 170 Oak Brook, IL 60523-4413 | 630-990-0001 | http://www.werc.org/ |

# APPENDIX B: General Dock Design

*This is an excerpt on* general *dock design from the publication* Time, Space, Cost Guide to Better Warehouse Design *by Maida Napolitano and the Staff of Gross & Associates, Copyright 1994, Alexander Research & Communications, Inc., pp. 33-38. Although many design and safety principles apply, the following text is not specific to cross docking. Designs may differ based on the type and throughput of the program. Docks for cross docking are typically deeper (60 to 70 feet in depth) with larger truck yards. For other specific guidelines, refer to Chapter 3, Sections 3.2 to 3.3.*

\* \* \*

A substantial part of warehousing operations is concentrated in the receiving and shipping area, or the *dock area*. Failure to plan carefully for your facility's shipping and receiving needs may result in inadequately sized areas, costly future renovations, and equipment problems. Most importantly, it may cost you demurrage for delaying your carriers and impede efficient shipping and receiving operations.

## Design Year Requirements

Receiving/Shipping operations have to be designed for the peak load estimates of the design year. The design year may be 5 to 7 years into the future. Design year estimates are often calculated by using historical data. Design year requirements are expressed in terms of expected:

- Frequency of receipts and shipments
    - —Number of trucks arriving
    - —Times of arrival
    - —Loading and unloading times
    - —Volume by product lines

**FIGURE 81**

**Product Flow Patterns**

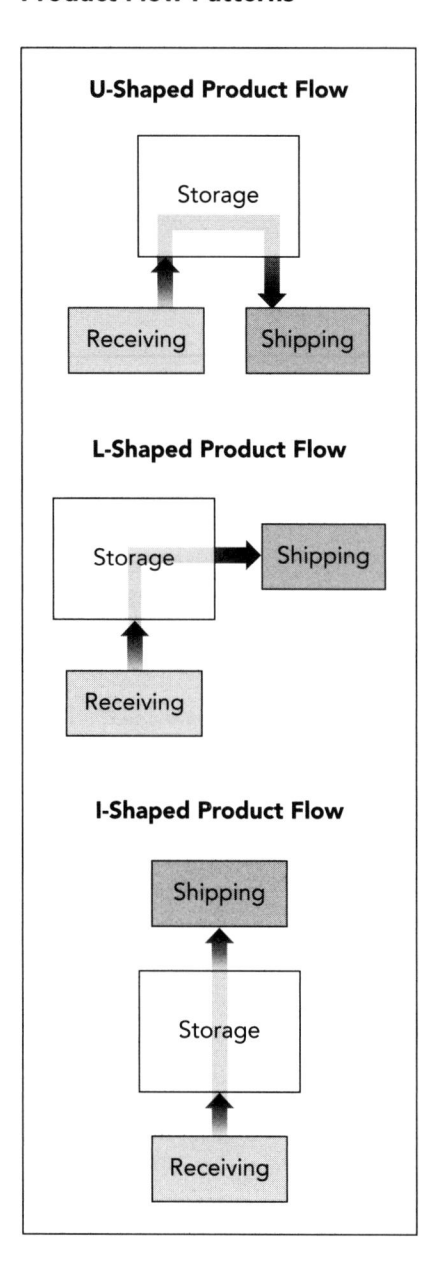

**U-Shaped Product Flow**

Storage
Receiving | Shipping

**L-Shaped Product Flow**

Storage → Shipping
Receiving

**I-Shaped Product Flow**

Shipping
Storage
Receiving

—Volume by mode of transportation

Other questions to be answered are:

- What is the number and location of suppliers who will ship to the facility?
- Which carriers will deliver/pick up?
- What modes of transportation will be used?
- What is the volume in cubic feet and typical shipment sizes in cubic feet for each supplier/receiver?
- Are loads unitized or hand stacked?

## Dock Design

The design of the dock area includes the following three major steps:

- determining the proper location of the docks,
- determining the number and size of dock doors required, and
- specifying the equipment that is essential to dock operations.

### Determining the Proper Location of the Docks

Depending on the layout of the facility, the receiving and shipping dock doors can be in the same area as in a U-shaped product flow or in different areas of the warehouse, as in an L-shaped or I-shaped product flow. Figure 81 illustrates these different flow patterns. Having the shipping and receiving doors close to each other allows for more flexibility for dock usage, promotes faster cross docking capability, and permits consolidation of the supervisory function for the two operations. In addition, less overall space may be required. Conversely, separating them may improve security and reduce congestion.

In some cases, space restrictions may be the overriding factor in deciding where to locate docks. However, when there are a number of choices possible, there are other practical factors which should be considered.

- It is safest and quickest for vehicles to approach the docks in a counterclockwise direction. In this direction, the driver can look out the driver's side window when backing up. For these same reasons, the distance between centers of the doors should be at least 12 feet 6 inches. The actual distance between the doors may be greater, depending upon other factors such as building column spacing and staging requirements.

- There must be sufficient apron space and a waiting area for trucks away from the dock doors. For 12 feet 6 inch door centers where trucks approach in a counterclockwise direction, the apron space should be approximately twice the length of the

APPENDIX B  *General Dock Design*   171

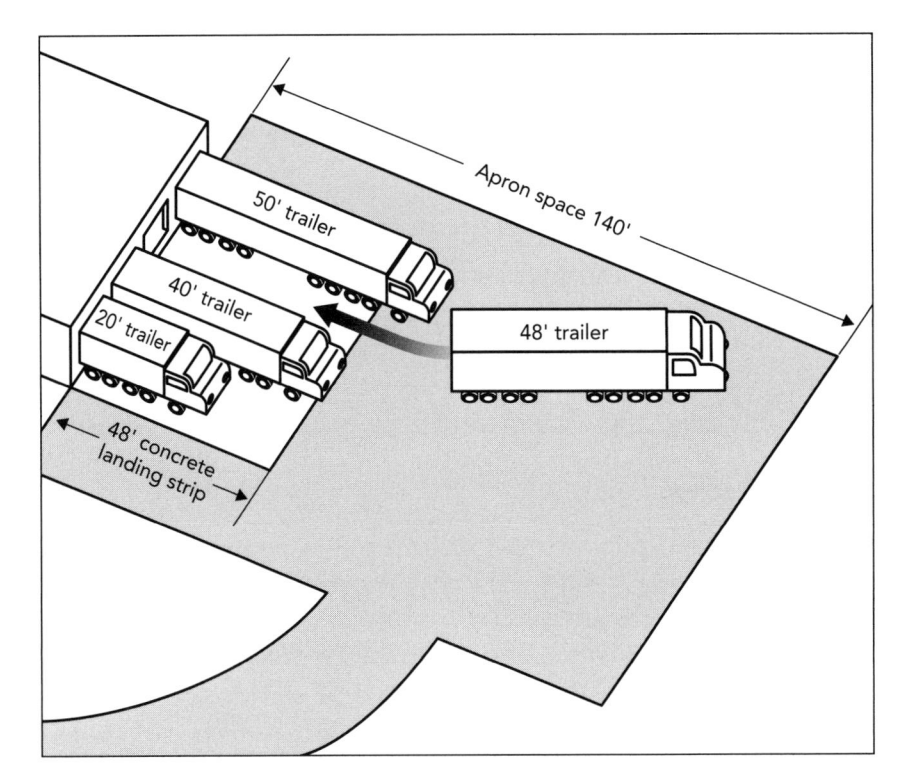

**FIGURE 82**

**Sample Dock Layout**

longest vehicle plus an additional 5 feet as a safety margin. As the distance between the doors increases, apron space can be reduced. For two-directional traffic, roads should be 23 feet wide. For one-way roads, the minimum width should be 12 feet. Gates and approaches to roadways should be at least 30 feet wide for two-directional traffic and at least 20 feet for one-way traffic.

- Some facilities incorporate a concrete landing strip into the driveway that is long enough to accommodate the longest trailer that the facility will serve. This will prevent the landing gear and kingpin jack from sinking into the asphalt in hot weather. Figure 82 illustrates a sample dock layout with counterclockwise vehicle approach, and relative sizes for landing strip and apron areas.

- Knowing something about the geography of the area where you plan to locate your facility can also be valuable. In a hilly area, your dock may require special equipment. It is preferable to have a level approach to the docks, but if a slope is unavoidable you will need to design bumpers, seals, and canopies to accommodate the angle of the truck's approach. If possible, the doors should face away from the prevailing winds. An important related consideration is the nature of the neighboring businesses. For example, if there are incinerators or large areas of garbage disposal on nearby property, and you plan to leave dock doors open, you should consider locating the doors on the opposite side of your building to prevent smoke, fumes, flies, or other undesirable objects from entering the facility.

**Number and Size of Dock Doors Required**

Knowledge of dock requirements of your operation is essential in determining the number of doors your facility will require. It is important to know the pattern of truck arrivals and departures, the length of time it takes to load or unload the trucks, and the total number of pallets to be handled in a typical day.

For example, if XYZ Company expects up to 20 carriers per 8-hour day and each carrier spends an average of 3 hours and 30 minutes at the facility, a door can only be used twice a day. Therefore, a minimum of 10 doors would be required. This example illustrates one of the simplest methods of calculating dock door requirements. Other increasingly complex and correspondingly more accurate formulas can be used when more detailed information is available.

*Door Width.* You should know the size of the trucks expected at the facility when you plan the dock height and door sizes. Most trailers currently in use are 96 inches wide, but most of the trailers being built today are 102 inches wide. Loading and unloading the rearmost pallets in a 102-inch-wide trailer can be difficult when the trailer is backed into a standard 96 inch wide dock door. Dock doors must be wide enough to accommodate both trailer sizes. We typically recommend 108-inch-wide doors. These allow a 102-inch-wide trailer to back up to the dock with some margin for error.

*Door Height.* Maximum allowable height on interstate highways is 13 feet 6 inches, although some states allow higher vehicles on local roads. With standard 48-inch-high docks, doors that are at least 9 feet 6 inches high allow full access to the trailer. Although most trailer beds are between 48 and 52 inches from the ground, new low profile truck tires have made possible "high cube" trailers that sit only 36, and in some cases, 30 inches off the ground. There are two methods for allowing warehouse vehicles to enter a truck lower than the dock.

The most common approach is to bring the dock to trailer level by use of a dock leveler. However, a dock leveler that can compensate for a 12- to 18-inch difference in dock and vehicle height would have to be at least 10 feet long. Another device, called a *dual dock*, also allows you to lower the truck to the level of the trailer bed.

The other approach is to bring the trailer to dock level. Portable wheel ramps are one possibility, but a safer, more flexible, and increasingly popular method is a trailer leveler, which raises the trailer to the dock level.

APPENDIX B   *General Dock Design*

The dock should be kept clear of material. At least 15 to 40 feet of clear dock space from the doors should be laid out for staging. A main aisle between the clear dock space and the staging area should be on the plan. For cross docking, new facilities are suggesting up to 70 feet of clear dock space.

## Other Dock Equipment Needs

There are various kinds of equipment used for moving material at the dock area. Following are some considerations for dock equipment:

- Lift trucks should be equipped with seat belts, backup alarms, horns, overhead guards, tilt indicators, and an on-board fire extinguisher. Drivers must be screened and trained to ensure safe driving practices.

- Pedestrian traffic should be restricted in the dock area with a clearly marked walkway. Guardrails should define the pedestrian walkway. Convex mirrors should be installed at blind corners.

- Vehicle restraint systems should be installed at the dock doors which will ensure that there is no accidental pull-away by a trailer while it is being serviced. Wheel chocks serve this purpose and are inexpensive, but they are not as reliable as trailer restraining devices, particularly in the snow. These restraints can be simple mechanical devices to sophisticated electronic ones with engagement warning lights for dock personnel and trailer drivers. They should be effective in all weather conditions. Dock personnel should be trained to visually verify that the vehicle restraints are always engaged.

- If wheel chocks are used, the drivers are responsible for placing them and the dock workers must check them.

- Dock levelers that will provide a gentle grade into trailers of all heights should be chosen. They should be able to service 8-feet-6-inch wide trailers. Leveler capacity must be adequate to handle all load weights. Dock levelers should have the following safety features: full range toe-guards, ramp free-fall protection, automatic recycling, safety stops, safety lip.

- Proper lighting is essential to fast, safe shipping/receiving operations and dock lights should be installed for use inside trailers. Trailer approach guide lights on both sides of the dock doors would be helpful as guides for night operations.

- Weather sealing is advised for operations in cold weather climates. It saves energy and increases safety by preventing rain, snow, dirt, and debris from blowing on the docks where they might cause slippage and lift truck skids. Placing canopies over docks also serves as protection from inclement weather.

**Safety on the Dock**

With its high concentration of activities, the dock is the area where most accidents are likely to happen. Dock accidents can become a significant human and cost factor if the proper equipment is not emphasized. Costs are both direct, such as compensation benefits and medical treatment, and indirect such as production losses, the value of spoiled or damaged products, and equipment repair. Proper training of new employees, proper selection and maintenance of equipment, and well-designed operating procedures are necessary for smooth shipping/receiving operations.

According to OSHA all dock employees must be required to attend a comprehensive dock training program. They must be trained on the use of all dock equipment (dock levelers, vehicle restraints, driving rules, etc.), and on safety/emergency procedures. Dock equipment must be maintained to ensure proper operation and to avoid malfunctions.

Reprinted with permission.

# Bibliography

Ackerman, Kenneth B., "The Many Flavors of Cross Docking," *Warehousing Forum*, September 1996.

Andel, Tom, "Efficient Transportation Starts in the Warehouse," *Integrated Warehousing & Distribution*, June 1998.

Beech, Jeff, "Crossdocking Meets Demand," *Food Logistics*, May 1998.

Bernard, John, "The Road to Successful Cross Docking: A Practitioner's Perspective," USF Logistics presentation, October 13, 1998.

Blaser, Jim, "Cross Docking *Can* Work in the Food Industry," *Consumer Goods Manufacturer*, n.d., pp. 1-9.

Casper, Carol, "Flow-through: Mirage or Reality," *Food Logistics*, October/November 1997.

Cockerham, Paul, "Throughput Rules Distribution," *Frozen Food Age*, July 1998, pp. 1, 21-22.

Cooke, James Aaron, "Cross-docking rediscovered," *Traffic Management*, November 1994.

Cooke, James Aaron, "Cross-Docking Software: Ready or Not?," *Logistics Management*, October 1997.

Cooke, James Aaron, "Do You Have What It Takes To Cross Dock?," *Logistics Management*, September 1996.

"Crossdocking in Supply Chain Flow," *Material Handling Engineering*, 1998/99.

"Cross Docking: Process for Success," Special Report for FMI/GMA Replenishment Excellence Conference, New Orleans, LA, September 1995.

Daugherty, Patricia J., "Strategic Information Linkage," *The Logistics Handbook*, New York: The Free Press, 1994.

Dell, W. Frank II, "A Case for Wholesale Change," *Supermarket Business*, October 1997.

Donaldson, Harvey, "Schedule-Driven Cross-Docking Networks," Research paper from the Georgia Institute of Technology, on the web at http://www.tli.gatech.edu/lms2000/

"ECR: From Theory to Practice," *Logistics*, 1994.

Eyestone, Dave, "Operations and Systems Tools for Stocking and Flow Through Distribution," *Transtech Consulting, Inc., Brochure*, Spring 1997.

"Flow-Through DC Yields Savings for Fred Meyer Inc.," *Chain Store Age*, New York, October 1995.

Garry, Michael, "A Blueprint of the Future," *Progressive Grocer*, October 1994.

Garry, Michael, "Meijer's Caveats," *Progressive Grocer*, May 1996.

Gilmore, Dan, "Transitions in Transportation," *IDSystems*, July 1996.

Grant, Eugene, et al., *Principles of Engineering Economy*, 7th ed., New York: John Wiley & Sons, 1982.

Gross & Associates, *Rules of Thumb*, A brochure that provides warehousing & distribution equipment costs and throughput standards. For copies: www.GrossAssociates.com.

Haedicke, Jack, "The ABCs of Profitability," *Progressive Grocer*, Jan. 1994.

Halverson, Richard, "Fred Meyer Defines QR Success," *Discount Store News*, April 17, 1995.

Harmon, Roy L., *Reinventing the Warehouse*, New York: The Free Press, 1993.

Harps, Leslie Hansen, "Crossdocking for Savings," *Inbound Logistics*, May 1996.

Harrington, Lisa, "New tools to automate your supply chain," *Transportation & Distribution*, December 1997.

Harrison, Dan, "10 Tips On Outsourcing," *Frozen Food Age*, December 1998.

"Internet transactions are here already: goodbye EDI?," *Frozen Food Age*, July 1999.

Jabbonsky, Larry, "Replenishment Logistics: The Pause That Refreshes Supermarket Distribution Costs," *Beverage World*, March 1993.

Jones, Bill, et al., "Merge in Transit Works: We Can Prove It," *Annual Conference Proceedings*, Council of Logistics Management, Oct. 8-11, 1995.

Knill, Bernie, "Information Pulls Food Distribution," *Material Handling Engineering*, July 1997.

Lear-Olimpi, Michael, "Looking for the fast lane," *Warehousing Management*, April 1999, pp. 26.

Lindeburg, Michael R., *Engineer in Training Review Manual*, California: Professional Publications, Inc., 1982.

Malloy, Amy, "Experience helps: E-shopping rises 270%," *Computerworld*, Jan. 10, 2000.

Martin, Christopher, "Integrating Logistics Strategy in the Corporate Financial Plan," The Logistics Handbook, New York: The Free Press, 1994.

McEvoy, Kevin, "DSD or cross-docking—or both?," *Progressive Grocer*, March 1997.

McLeod, Marcia, "Cutting Both Ways," *Supply Management*, London, April 1, 1999.

Menezes, Joaquim, "ERP's future lies in supply chain," *Computing Canada*, April 16, 1999.

Minahan, Tim, "Are Buyers Gumming Up The Supply Chain?," *Purchasing*, January 1997.

Minahan, Tim, "Dell Computer sees suppliers as key to JIT," *Purchasing*, September 4, 1997.

Minahan, Tim, "Miller SQA tweaks JIT system for quick response," *Purchasing*, September 4, 1997.

Moore, Bert, "Where do you want to know today," *Automatic I.D. News*, August 1999.

Muller, E J, "Faster, faster, I need it now!" *Distribution*, February 1994.

Napolitano, Maida, *Using Modeling To Solve Warehousing Problems*, Illinois: Warehousing Education and Research Council, 1998.

Nelson, James A., "Inventory in Motion: Adding Value Through Cross Docking," Presentation by Ryder Integrated Logistics for Warehousing Education and Research Council Conference, April 1999.

"Not As Simple As It Looks," *Distribution Center Management*, " February 1998.

Novack, Robert, et al., *Creating Logistics Value: Themes for the Future*, Illinois: Council of Logistics Management, 1995.

Patterson, Donald, "Pausing-In-Transit: A Distinctive Option in Distribution." *Warehousing Forum*, The Ackerman Company, April 1999.

Pohlen, Terrance, "Applications of Activity-Based Costing Within Logistics: Who is Using Activity-Based Costing and Where?," *Annual Conference Proceedings*, Council of Logistics Management, October 16-19, 1994.

Ratliff, H. Donald, et al., "Network Design for Load-Driven Cross Docking Systems," Research paper from the Georgia Institute of Technology, on the web at http://www.tli.gatech.edu/lms2000/

Retrotech, Inc. , "Ahead of the Rest With Cross Docking," Brochure provided for their ACTIV systems. Contact Retrotech Incorporated, 610 Fishers Run, P.O. Box 586, Fishers, NY 14453-05786, (716) 924-6330.

Rouland, Renee Covino, "Perpetual Partners," *Discount Merchandiser*, December 1994.

Schaffer, Burt, "Cross Docking Can Increase Efficiency," *Automatic I.D. News*, July 1998.

Schaffer, Burt, "Implementing a Successful Cross Dock Operation," *IIE Solutions,* October 1997.

Schaffer, Burt, "Plan Before You Cross Dock: How to Implement a Successful Cross Docking Operation," Tompkins Associates presentation to Warehousing Education and Research Council Annual Conference, April 1999.

Schwartz, Beth M., "Balance Warehouse & Market Need," *Transportation & Distribution*, September 1998.

# BIBLIOGRAPHY

Schwind, Gene, "A Systems Approach to Docks and Cross Docking," *Material Handling Engineering*, February 1996.

Shamlaty, Ron, "This is not your father's warehouse," *IIE Solutions*, January 2000, pp. 31-36.

Sherman, Richard, "From vision into action," *Beverage World*, September 1995.

Small, Stefanie, and McAndrews, Maureen, "Keeping Inventory Moving," *Warehouse Management*, September 1998.

"Speeding Through the Warehouse," *WERCSheet*, July/August 1999.

Speh, Thomas W., PhD., *A model for determining total warehousing costs: For private, public and contract warehouses*, Illinois: Warehousing Education and Research Council, 1990.

"Sports Authority Eliminates Storage Space and Processes Most Orders Within 24 to 36 Hours," *PkMSolution News*, Spring 1999.

Stein, Tom, "Optum lets users peek into logistics," *Informationweek*, May 3, 1999.

"Step into the Pool," *Transportation & Distribution*, August 1996.

Stephens, Bill, "Roads smoothes out for NS auto hubs," *Trains*, June 1998, pp. 21-22.

"Streamlining Apparel Distribution," *Apparel Industry Magazine*, March 1994.

Symons, Allene, "Contract Negotiations Aim for Win-Win Relationships," *Drug Store News*, December 1998.

Thayer, Warren, "Logistics: Life in the Fast Lane," *Frozen Food Age*, July 1999.

"The flow-through concept: We don't store it—we ship it," *Modern Materials Handling*, June 1990.

"Tips, Tactics, and Strategies for Starting a Cross-Docking Process," *Inventory Reduction Report*, March 1998.

Tompkins, James, et al., *Facility Planning*, 2nd ed., New York: John Wiley & Sons, 1996.

Trebuchon, Maurice, and Muskett, David, "Cross Docking Challenges," Presentation at Council of Logistics Management, October 1997.

Wagar, Kenneth, "Cross Dock and Flow Through Logistics for the Food Industry," *Annual Conference Proceedings*, San Diego, CA: Council of Logistics Management, October 1995.

Wagar, Kenneth, "The Logic of Flow-Through Logistics," *Supermarket Business*, June 1995.

Weinstein, Steve, "An Efficient Combination?," *Progressive Grocer*, December 1994.

Westburgh, Jesse, "Cross Docking in the Warehouse—An Operator's View," *Warehousing Forum*, August 1995.

White, John, III, *Cross Docking Principles and Systems*, The Warehousing Short Course, September 1996.

Wurz, Al, "Automatic Data Collection," *Transportation & Distribution*, November 1994.

# Index

ABC, 40, 48, 69, 116–121, 119, 143, 176
Ackerman, Kenneth, vii, 175
ACTIV Systems, 70, 176
Activity Based Costing. *See* ABC
Advance Shipping Notices. *See* ASN
American Freightways, 46
Andel, Tom, 66, 175
ANSI, 142
Apron space, 170–171
ASN, 53, 56, 63, 64, 73, 81, 86–88,
Assortments, 13, 59, 62, 69, 108

Backorders, 40
Bar coding, 22, 31, 44, 48, 51, 53, 54, 56, 57, 63, 72, 80, 107
Beech, Jeff, 175
Bernard, John, 175
BJ's, 13
Blaser, Jim, 66, 102, 175

CAO, 23, 31, 53, 57, 161,
Carmon, Virginia, vii, 28, 63
Casper, Carol, 2, 69, 175
Catalyst International, vii, 40
Category management, 17, 22, 31
Chargebacks, 64
Coca-Cola™ Retailing Research Council, 102
Coca-Cola™ Co., 119
Cockerham, Paul, 66, 175
Commerce Net, 24
Computer-Assisted Ordering. *See* CAO
Consultants, 97, 114, 115, 125–126, 142
Consulting Services Company, 103,
Contingency plans, 128, 133–134
Continuous Replenishment Programs. *See* CRP
Converting, 133
Conveyors, 13, 71, 76-78, 81–82, 85, 97, 129, 130, 135, 148, 152, 156, 161
Cooke, James Aaron, 26, 66, 103, 175
Copacino, William, 165

Cost model, 20, 41, 107-121, 142
Costco, 13
Council of Logistics Management, 26, 66, 102, 103, 121, 165, 166, 176, 177
Cross docking
 Benefits, drawbacks, 19
 Distributor, 12
 Ideal conditions, 14-16
 Initiator, 66, 90
 Manufacturing, 9-12
 Opportunistic, 14
 Retail, 13-14, 72, 80
 Transportation, 12-13
CRP, 22–24, 31
Cube movement, 33, 34, 35–36, 37

Damage, 18, 19, 43, 90, 112, 113, 119
Daugherty, Patricia, 66, 175
Dell Computer Corp., 13, 39–40, 176
Dell, W. Frank II, 175
Demurrage, 95, 102, 169
Design year, 60, 93, 138, 169–170
Direct from manufacturer. *See* Warehouse bypass
Direct product profitability (DPP), 41, 116–117, 121
Direct store delivery. *See* DSD, Warehouse bypass
Display-ready products, 13, 40, 48, 57, 59, 64, 69, 148
Distributor, 12, 13, 16, 57, 31, 143
Dock design, general, 142, 169-174
Dock door height, 172
Dock door width, 172
Dock leveler, 172
Donaldson, Harvey, 175
Door-per-store, 53, 74, 82, 95, 132, 149
Drop ship. See Warehouse bypass
DSD, 18, 26, 65, 106, 176
Dual dock, 172

e-commerce, 23–25, 46
ECR, 22–23, 31, 40, 60, 175

EDI, 22–24, 31, 44, 45–46, 48–49, 51, 53, 54–57, 63, 72, 81, 86–88, 107, 133, 142, 148, 176
Efficient Consumer Response. *See* ECR
Electronic commerce. *See* e-commerce
Enterprise Resource Planning (ERP), 25
Equivalent uniform annual cost (EUAC), 100–101, 103, 115–116, 159, 163
Exel Logistics, 90
Expediting, 8
Eyestone, Dave, 175

Facility location, 125, 144
Federal Express, 8, 12, 24
FIFO Granularity, 39
First In/First Out (FIFO), 39
Floor ready. *See* Display-ready products
Flow through, 3, 22, 40, 66, 102, 175, 177
Food Distributors International Productivity Analysis, 22
Ford Automotive, 3, 136
Fred Meyer, 3, 175
Freight houses, 8

Gainsharing, 62
GANTT chart, 127, 128, 144, 165
Garry, Michael, 14, 145, 175
GATX Logistics, 4–5
General administration (GA) costs, 109–110
Gilmore, Dan, 175
Gordon, Michele, 165
Grant, Eugene, 103, 175
Grocery industry, 31, 55, 102, 166
Grocery pilot programs, 18–19
Gross & Associates, ii, viii, ix, 102, 147, 169, 175

Haedicke, Jack, 119, 121, 175
Halverson, Richard, 3, 175
Handling expenses, 111, 112, 113
Hannaford Bros., 2
Harmon, Roy, 175
Harps, Leslie Hansen, 12, 175
Harrington, Lisa, 26, 176
Harrison, Dan, 176
Hartz Mountain Corporation, 63, 64
Herman-Miller Inc., 9
Hills Department Stores, 63, 64
Home Shopping Network Inc., 25

Information systems, 6, 10, 15–16, 19, 20, 21, 22, 24, 25, 26, 30, 31, 43, 45, 46, 47, 48, 49, 50–51, 52, 53, 54, 55, 57, 59, 66, 67, 68, 69, 71, 72–74, 77, 79–81, 83, 85, 86–88, 89, 91, 93, 96–97, 114–115, 124–125, 127, 128, 129, 131, 136–137, 138, 142, 148–149, 156–161, 163, 165, 175–176
Internet, 23–25, 46, 53, 54, 66, 176
Interstore transfers, 41

Jabbonsky, Larry, 176
JIT, 3, 4, 9–11, 16, 19, 22, 40, 48, 51, 83, 91, 126, 176
Jones, Bill, 26, 176
Just-In-Time. *See* JIT

Kinetic Computer Corp., 46
Kmart, 9
Knill, Bernie, 26, 176

Label compliance, 44, 46–47
Last In/First Out (LIFO), 39
Lear-Olimpi, Michael, 12, 13, 46, 176
Less-than-truckload. *See* LTL
Lever Bros., 69
Lewis, James, 165
Lift trucks, 32, 38, 55, 71–73, 76–78, 85, 94, 96–97, 110, 135, 148, 152, 156, 158, 159, 173–174
Lindeburg, Michael, 103, 176
Lot control, 39
LTL, 46, 149

Malloy, Amy, 26, 176
Martin, Christopher, 121, 117, 176
Mass merchandisers, 9, 13, 64–65, 66
MaxiCode, 47
McAndrews, Maureen, 144, 177
McEvoy, Kevin, 26, 176

McLeod, Marcia, 103, 176
Meijer, 13–14, 175
Menezes, Joaquim, 26, 176
Menlo Logistics, 9–11
Mercer Management Consulting, 102
Merge-in-transit, 13, 26, 176
Miller Brewing Co., 11, 133, 176
Miller SQA, Inc., 9–10, 176
Minahan, Tim, 11, 40, 176
Minimum Attractive Rate of Return (MARR), 100, 101, 115, 116, 160
Mitsubishi Motor Manufacturing of America, 4–5
Mixing centers, 3–4, 136
Modes of transportation, 3, 170
Montgomery Ward, 8
Moore, Bert, 145, 176
Moore, Thomas, 11, 26
Muller, E J, 176
Muskett, David, 177

Napolitano, Maida, ix, 66, 98, 165, 169, 176
Nelson, James, 176
Norek, Christopher, 66
Norfolk Southern, 4, 136
Novack, Robert, 66, 121, 176

Obsolescence, 18, 19, 112, 124
Occupational Safety and Health Association (OSHA), 174
Ohio State University, 121
Operating administration (OA) costs, 109–110
Optum, Inc., 25, 26, 177
Order completion profile, 33, 35, 37
Oshawa Foods, 93

Packaging, 39, 71, 166
Pallet jacks, 32, 38, 55, 71, 72, 77, 78, 84, 85, 94, 97, 110, 135, 148, 152, 154, 156, 158, 159
Pallet license plates, 46, 53, 56, 57
Parcel delivery companies, 12
Patterson, Donald, 102, 176
Payback period, 99, 159–160
PDF417, 47
Performance specification bid, 134, 135
Perishability, 39
Pick-to-pallet, 54, 84, 86, 94
Pilferage, 18, 19, 112

Pilot program, 18, 19, 20, 21, 28, 124, 127, 134, 135, 137–139, 140, 144, 161
Planning horizon, 30, 59, 60, 61, 66, 93, 97, 116
Pohlen, Terrance, 121, 176
Point of allocation, 68
Point of consolidation, 68
Point of Sale. *See* POS
Pool car, 8
POS, 16, 22–23, 31, 40, 53, 57, 63, 88, 161
Post-allocated cross dock operator consolidation, 68, 69, 83-86, 87, 88, 92, 93, 147
Pre-allocated cross dock operator consolidation, 68, 69, 71, 73, 76, 80, 85, 89, 90, 91, 93, 94, 107, 149
Pre-allocated supplier consolidation, 3, 40, 63, 68, 71, 85, 89, 91, 93, 94, 107, 120, 132, 149, 163
Present worth (PW), 100–101, 103, 115
Procter & Gamble (P&G), 63
Product specification plans, 132–133
Programmable Logic Controllers (PLC), 78
Pull distribution, 22, 163

QR, 3, 22, 31, 40, 60, 175, 176
Quick Response. *See* QR

Radio frequency, 11, 51, 53, 54, 56, 58, 72, 84, 85, 86, 87, 96, 97, 107, 148, 152
Ramps, 3–4, 172
Rand McNally Logistics Solutions, 12
Rapistan Systems, 77, 79, 102
Rate of return (ROR), 100–101, 116
Ratliff, H. Donald, 4, 176
Retrotech, Inc., 70, 120, 176
Return on Investment (ROI), 16, 20, 63, 100, 107, 114, 115–116
Richfood, 141–142
Risk assessment, 101
Robeson, James, 165
Rouland, Renee, 63, 176

Salvage value, 100, 103, 160
Sam's Club, 13
SCE, 25-26
Schaffer, Burt, 66, 102, 176
Schwartz, Beth, 176
Schwind, Gene, 26, 177
Scissor lifts, 78, 97